edexcel
advancing learning, changing lives

Mechanics 1

Edexcel AS and A level Modular Mathematics

Susan Hooker
Michael Jennings
Bronwen Moran
Laurence Pateman

Contents

About this book

This book is designed to provide you with the best preparation possible for your Edexcel M1 unit examination:

- The LiveText CD-ROM in the back of the book contains even more resources to support you through the unit.
- A matching M1 revision guide is also available.

Finding your way around the book

Brief chapter overview and 'links' to underline the importance of mathematics: to the real world, to your study of further units and to your career

Every few chapters, a review exercise helps you consolidate your learning

Detailed contents list shows which parts of the M1 specification are covered in each section

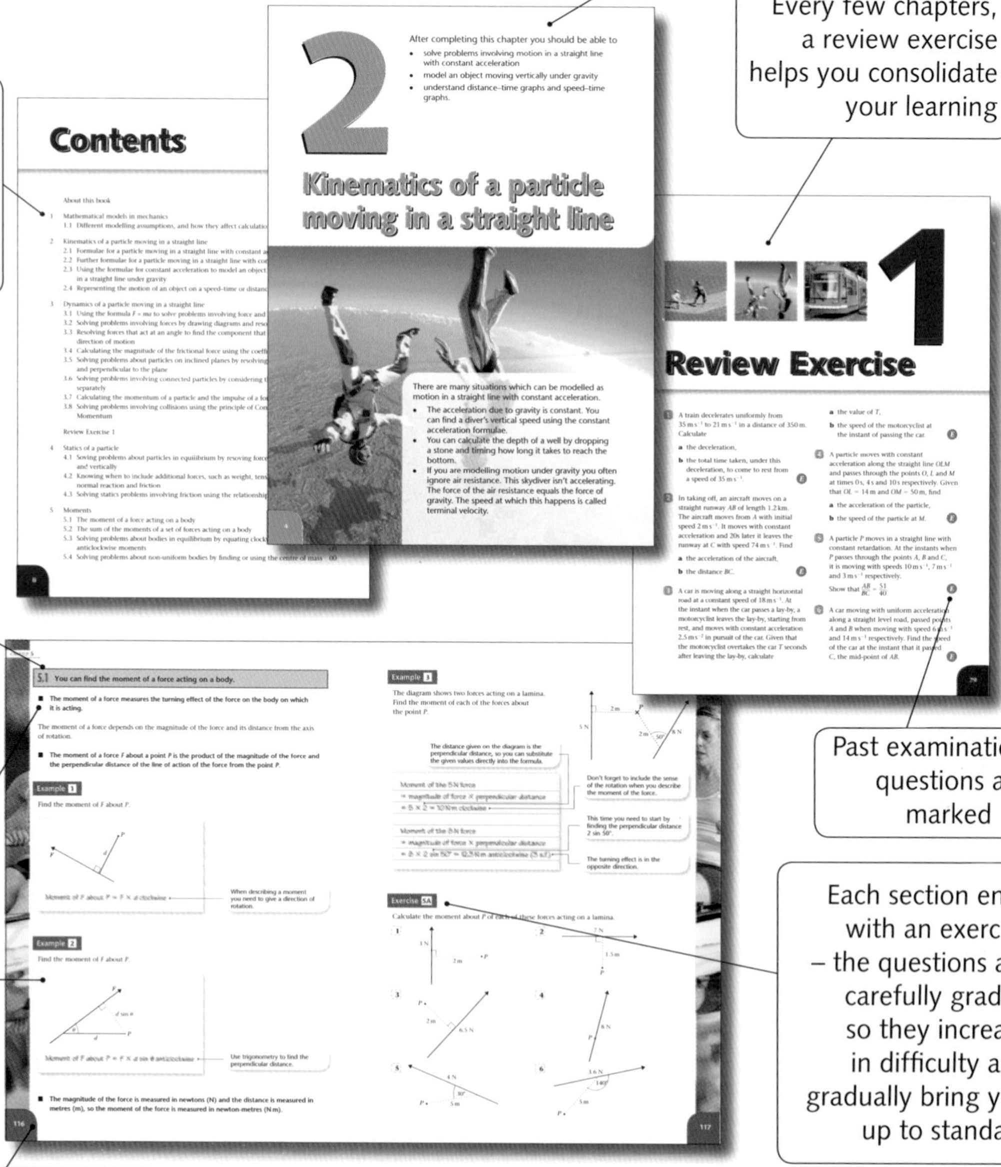

Each section begins with a statement of what is covered in the section

Concise learning points

Step-by-step worked examples

Past examination questions are marked 'E'

Each section ends with an exercise – the questions are carefully graded so they increase in difficulty and gradually bring you up to standard

Each chapter has a different colour scheme, to help you find the right chapter quickly

Each chapter ends with a mixed exercise and a summary of key points.

At the end of the book there is an examination-style paper.

LiveText software

The LiveText software gives you additional resources: Solutionbank and Exam café. Simply turn the pages of the electronic book to the page you need, and explore!

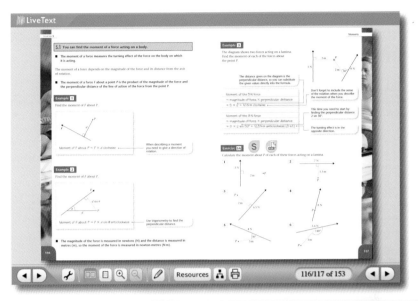

Unique Exam café feature:

- Relax and prepare – revision planner; hints and tips; common mistakes
- Refresh your memory – revision checklist; language of the examination; glossary
- Get the result! – fully worked examination-style paper

Solutionbank Edexcel AS and A Level Modular Mathematics

Mechanics 1   1 Mathematical models in Mechanics

Exercises: A ▾   Questions: 2 ▾

Question:

Describe briefly the process of refining a mathematical model.

Solution:

Predictions based on the model are compared with observed data.

In the light of this comparison the model may be adjusted (refined).

The process of collecting observed data and comparing with revised prediction from the model is repeated.

Solutionbank

- Hints and solutions to every question in the textbook
- Solutions and commentary for all review exercises and the practice examination paper

Pearson Education Limited, a company incorporated in England and Wales, having its registered office at 80 Strand, London WC2R 0RL. Registered company number: 872828

Text © Susan Hooker, Michael Jennings, Bronwen Moran, Laurence Pateman 2008

16
14

British Library Cataloguing in Publication Data is available from the British Library on request.

ISBN 978 0 435519 162

Edited by Susan Gardner
Typeset by Tech-Set Ltd
Illustrated by Tech-Set Ltd
Cover design by Christopher Howson
Picture research by Chrissie Martin
Index by Indexing Specialists (UK) Ltd
Cover photo/illustration © Science Photo Library / Laguna Design
Printed in China (CTPS/14)

Acknowledgements

The author and publisher would like to thank the following individuals and organisations for permission to reproduce photographs:

iStockPhoto.com / Amanda Rohde p**1**; Alamy / Andi Duff p**4**; iStockPhoto / gocosmonaut p**37**, Alamy / Corbis Premium RF p**92**; Rex Features / Aurora Photos p**116**; Corbis / Bob Sacha p**133**

Every effort has been made to contact copyright holders of material reproduced in this book. Any omissions will be rectified in subsequent printings if notice is given to the publishers.

Disclaimer

This Edexcel publication offers high-quality support for the delivery of Edexcel qualifications.

Edexcel endorsement does not mean that this material is essential to achieve any Edexcel qualifications, nor does it mean that this is the only suitable material available to support any Edexcel qualification. No endorsed material will be used verbatim in setting any Edexcel examination/assessment and any resource lists produced by Edexcel include this and other appropriate texts.

Copies of official specifications for all Edexcel qualifications may be found on the Edexcel website – www.Edexcel.com.

After completing this chapter you should be able to

- understand the significance of different modelling assumptions, and how they affect the calculations in a particular problem.

Mathematical models in mechanics

Mechanics is the branch of mathematics which deals with the action of forces on objects. Mechanics can be used to answer questions about many familiar situations – the motion of cars, the speed of a parachutist, the stresses in a bridge or the motion of the Earth around the Sun.

There are many factors that can complicate real-world problems. The motion of a cricket ball might be affected by the spin of the ball, the effects of air-resistance or the roughness of the pitch. We can simplify a problem by creating a **mathematical model**. This model will involve making a number of **modelling assumptions**, such as ignoring air-resistance, or treating a three-dimensional object as a particle.

Modelling assumptions
- ignore air resistance
- perfect bounce on a flat pitch
- constant force due to gravity
- model ball as a particle

$g = 9.8 \, \mathrm{m \, s^{-2}}$

Modelling assumptions can simplify a problem, and allow us to carry out an analysis of a real-life situation using known mathematical techniques. Having fewer modelling assumptions will usually make a problem more difficult mathematically.

By modelling the motion of this cricket ball, we can use mathematical techniques to predict whether it will hit the stumps.

1.1 **You need to understand the significance of different modelling assumptions, and how they affect the calculations in a particular problem.**

Here are some common models and modelling assumptions that you need to know.

- **Particle** – an object which is small in comparison with other sizes or lengths can be modelled as a **particle**. This means that the mass of the object can be considered to be concentrated at a single point (a particle is often referred to as a point-mass). The fact that a particle has no dimensions means that we can ignore the rotational effect of any forces that are acting on it as well as any effects due to air resistance.

- **Rod** – an object with one dimension small in comparison with another (such as a metre ruler or a beam) can be modelled as a **rod**. This means that the mass of the object can be considered to be distributed along a straight line. A rod has no thickness (it is one-dimensional) and is rigid (it does not bend or buckle).

- **Lamina** – An object with one dimension (its thickness) very small in comparison with the other two (its length and width) can be modelled as a **lamina**. This means that the mass of the object can be considered to be distributed across a flat surface. A lamina has no thickness (it is two-dimensional). For example, a sheet of paper or metal could be modelled as a lamina.

- **Uniform body** – If an object is uniform then its mass is evenly distributed over its entire volume. This means that the mass of the body can be considered to be concentrated at a single point (known as the **centre of mass**), at the 'geometrical centre' of the body. For example, an unsharpened pencil could be modelled as a uniform rod. However, once it is sharpened then its centre of mass would not be at its mid-point and we would model it as a non-uniform rod.

- **Light object** – If the mass of an object is very small in comparison with the masses of other objects, we can model it as being **light**. This means that we can ignore its mass altogether and treat it as having zero mass. Strings and pulleys are often modelled as being light.

- **Inextensible string** – If a string does not stretch under a load it is **inextensible** or **inelastic**. In M1 you will model all strings as being inextensible.

- **Smooth surface** – If we want to ignore the effects of friction, we can model a surface as being **smooth**. This means that we assume there is no friction between the surface and any object which is moving or tending to move along it.

- **Rough surface** – If a surface is not smooth it is said to be **rough**. We need to consider the friction between the surface and an object moving or tending to move along it. For example, a ski slope might be modelled as a smooth or a rough surface depending on the problem to be solved.

- **Wire** – A rigid thin length of metal, which is treated as being one-dimensional, is referred to as a **wire**. A wire can be smooth or rough. We often consider beads which are threaded on a wire.

- **Smooth and light pulley** – In M1 you will model all pulleys as being **smooth** (there is no friction at the bearing of the pulley or between the pulley and the string) and **light** (the pulley has zero mass).

- **Bead** – A particle which can be threaded onto, and move freely along, a wire or string is called a **bead**.

- **Peg** – A support from which an object can be suspended or on which an object can rest is called a **peg**. A peg is treated as being dimensionless (it is treated as a point) and is usually fixed. A peg can be rough or smooth.

- **Air resistance** – When an object moves through the air it experiences a resistance due to friction. In M1 you will model air resistance as being negligible.

- **Wind** – Unless it is specifically mentioned, you can usually ignore any effects due to the wind in your models.

- **Gravity** – The force of attraction between all objects with mass is called **gravity**. Because the mass of the Earth is large, we can usually assume that all objects are attracted towards the Earth (ignoring any force of attraction between the objects themselves). We usually model the force of the Earth's gravity as uniform, and acting vertically downwards. The acceleration due to gravity is denoted by g and is always assumed to be constant at 9.8 m s^{-2}. This value is given on the front of the exam paper.

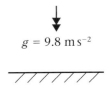

2

After completing this chapter you should be able to

- solve problems involving motion in a straight line with constant acceleration
- model an object moving vertically under gravity
- understand distance–time graphs and speed–time graphs.

Kinematics of a particle moving in a straight line

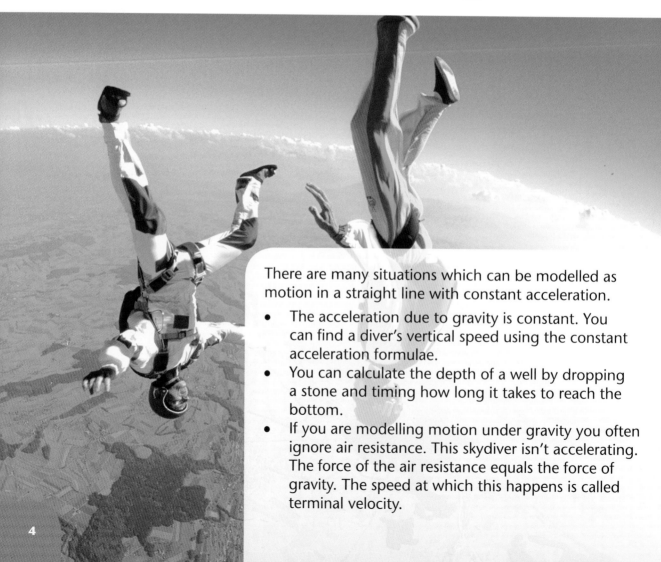

There are many situations which can be modelled as motion in a straight line with constant acceleration.

- The acceleration due to gravity is constant. You can find a diver's vertical speed using the constant acceleration formulae.
- You can calculate the depth of a well by dropping a stone and timing how long it takes to reach the bottom.
- If you are modelling motion under gravity you often ignore air resistance. This skydiver isn't accelerating. The force of the air resistance equals the force of gravity. The speed at which this happens is called terminal velocity.

2.1 You can use the formulae $v = u + at$ and $s = \left(\dfrac{u + v}{2}\right)t$ for a particle moving in a straight line with constant acceleration.

■ You need to learn this list of symbols and what they represent.

s	displacement (distance)
u	starting (initial) velocity
v	final velocity
a	acceleration
t	time

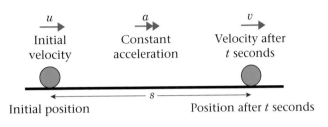

Displacements, velocities and accelerations have directions as well as sizes (or magnitudes). You can derive the formulae for motion in a straight line with constant acceleration.

- acceleration = $\dfrac{\text{change in velocity}}{\text{change in time}}$

  $$a = \dfrac{v - u}{t}$$

  $$\boldsymbol{v = u + at}$$

- distance moved = average speed × time

  average speed = $\dfrac{u + v}{2}$

  $$\boldsymbol{s = \left(\dfrac{u + v}{2}\right)t}$$

For an object moving horizontally, the positive direction is usually taken as left to right. The starting point of an object is usually taken as the origin from which displacements are measured.

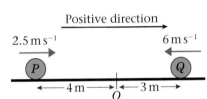

For P: $s = -4\,\text{m}$ and $v = 2.5\,\text{m s}^{-1}$

For Q: $s = 3\,\text{m}$ and $v = -6\,\text{m s}^{-1}$

Example **1**

A particle is moving in a straight line from A to B with constant acceleration $3\,\text{m s}^{-2}$. Its speed at A is $2\,\text{m s}^{-1}$ and it takes 8 seconds to move from A to B. Find **a** the speed of the particle at B, **b** the distance from A to B.

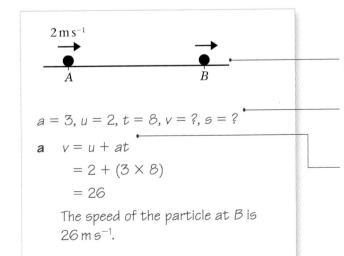

$2\,\text{m s}^{-1}$

A B

Start by drawing a diagram.

$a = 3,\ u = 2,\ t = 8,\ v = ?,\ s = ?$

Write down the values you know and the values you need to find.

a $v = u + at$

$= 2 + (3 \times 8)$

$= 26$

The speed of the particle at B is $26\,\text{m s}^{-1}$.

You need v and you know u, a and t so you can use $v = u + at$.

b $s = \left(\dfrac{u + v}{2}\right)t$ ─────────── Choose the right formula then substitute in
the values you know.

$\quad = \left(\dfrac{2 + 26}{2}\right) \times 8$

$\quad = 112$

The distance from A to B is 112 m.

Example 2

A cyclist is travelling along a straight road. She accelerates at a constant rate from a speed of
$4\,\mathrm{m\,s^{-1}}$ to a speed of $7.5\,\mathrm{m\,s^{-1}}$ in 40 seconds. Find **a** the distance she travels in these 40 seconds,
b her acceleration in these 40 seconds.

$4\,\mathrm{m\,s^{-1}} \qquad\qquad 7.5\,\mathrm{m\,s^{-1}}$

─────────── Model the cyclist as a particle.

$u = 4,\ v = 7.5,\ t = 40,\ s = ?,\ a = ?$

a $\quad s = \left(\dfrac{u + v}{2}\right)t$

$\quad\quad = \left(\dfrac{4 + 7.5}{2}\right) \times 40$

$\quad\quad = 230$

The distance the cyclist travels
is 230 m.

You need a and you know v, u and t so you
can use $v = u + at$.

b $\quad v = u + at$ ─────────── Substitute the values you know into the
formula. You can solve this equation to find a.

$7.5 = 4 + 40a$

$a = \dfrac{7.5 - 4}{40} = 0.0875$

You could rearrange the formula before you
substitute the values:

$a = \dfrac{v - u}{t}$

The acceleration of the cyclist
is $0.0875\,\mathrm{m\,s^{-2}}$.

In real-life situations values for the
acceleration are often quite small.
Large accelerations feel unpleasant and may
be dangerous.

■ **If a particle is slowing down it has a negative acceleration. This is called deceleration or retardation.**

Example 3

A particle moves in a straight line from a point A to a point B with constant deceleration $1.5\,\mathrm{m\,s^{-2}}$. The speed of the particle at A is $8\,\mathrm{m\,s^{-1}}$ and the speed of the particle at B is $2\,\mathrm{m\,s^{-1}}$. Find **a** the time taken for the particle to move from A to B, **b** the distance from A to B.

After reaching B the particle continues to move along the straight line with constant deceleration $1.5\,\mathrm{m\,s^{-2}}$. The particle is at the point C 6 seconds after passing through the point A. Find **c** the velocity of the particle at C, **d** the distance from A to C.

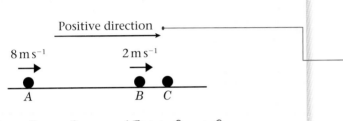

Positive direction

$8\,\mathrm{m\,s^{-1}}$ $2\,\mathrm{m\,s^{-1}}$

A B C

> Mark the positive direction on your diagram.

$u = 8, v = 2, a = -1.5, t = ?, s = ?$

> The particle is decelerating, so the value of a is negative.

a $v = u + at$

$2 = 8 - 1.5t$

$1.5t = 8 - 2$

$t = \dfrac{8 - 2}{1.5} = 4$

The time taken to move from A to B is $4\,\mathrm{s}$.

> You can use your answer from part **a** as the value of t.

b $s = \left(\dfrac{u + v}{2}\right)t$

$= \left(\dfrac{8 + 2}{2}\right) \times 4 = 20$

The distance from A to B is $20\,\mathrm{m}$.

c $u = 8, a = -1.5, t = 6, v = ?$

$v = u + at$

$= 8 + (-1.5) \times 6$

$= 8 - 9 = -1$

> The velocity at C is negative. This means that the particle is moving from right to left.

The velocity of the particle is $1\,\mathrm{m\,s^{-1}}$ in the direction $\overrightarrow{BA}$.

> Remember that to specify a velocity it is necessary to give speed and direction.

d $s = \left(\dfrac{u + v}{2}\right)t$

$= \left(\dfrac{8 + (-1)}{2}\right) \times 6$

> Make sure you use the correct sign when substituting a negative value into a formula.

The distance from A to C is $21\,\mathrm{m}$.

■ Convert all your measurements into base SI units before substituting values into the formulae.

Measurement	SI unit
time (t)	seconds (s)
displacement (s)	metres (m)
velocity (v or u)	metres per second (m s^{-1})
acceleration (a)	metres per second per second (m s^{-2})

Example 4

A car moves from traffic lights along a straight road with constant acceleration. The car starts from rest at the traffic lights and 30 second later the car passes a speed-trap where it is registered as travelling at 45 km h^{-1}. Find **a** the acceleration of the car, **b** the distance between the traffic lights and the speed-trap.

$45 \text{ km h}^{-1} = 45 \times 1000/3600 \text{ m s}^{-1}$

$\qquad\qquad = 12.5 \text{ m s}^{-1}$

Convert into SI units, using:
1 km = 1000 m
1 hour = 60 × 60 s = 3600 s

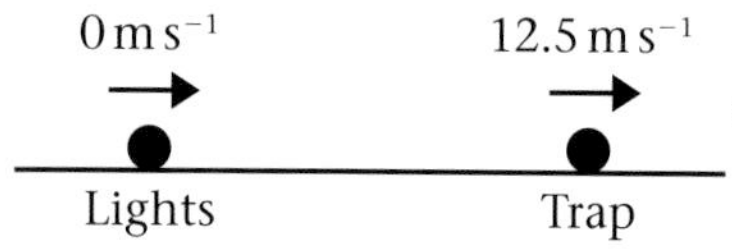

$0 \text{ m s}^{-1} \qquad\qquad 12.5 \text{ m s}^{-1}$

Lights   Trap

Model the car as a particle and draw a diagram.

$u = 0, v = 12.5, t = 30, a = ?, s = ?$

a $v = u + at$

$12.5 = 0 + 30a$

$a = \dfrac{12.5}{30} = \dfrac{5}{12}$

The acceleration of the car

is $\dfrac{5}{12} \text{ m s}^{-2}$.

The car starts from rest, so the initial velocity is zero.

This is an exact answer. If you want to give an answer using decimals, you should round to three significant figures.

b $s = \left(\dfrac{u + v}{2}\right)t$

$= \left(\dfrac{0 + 12.5}{2}\right) \times 30$

$= 187.5$

The distance between the traffic lights and the speed trap is 187.5 m.

Exercise 2A

1. A particle is moving in a straight line with constant acceleration $3 \, m \, s^{-2}$. At time $t = 0$, the speed of the particle is $2 \, m \, s^{-1}$. Find the speed of the particle at time $t = 6 \, s$.

2. A particle is moving in a straight line with constant acceleration. The particle passes a point with speed $1.2 \, m \, s^{-1}$. Four seconds later the particle has speed $7.6 \, m \, s^{-1}$. Find the acceleration of the particle.

3. A car is approaching traffic lights. The car is travelling with speed $10 \, m \, s^{-1}$. The driver applies the brakes to the car and the car comes to rest with constant deceleration in $16 \, s$. Modelling the car as a particle, find the deceleration of the car.

4. A particle moves in a straight line from a point A to point B with constant acceleration. The particle passes A with speed $2.4 \, m \, s^{-1}$. The particle passes B with speed $8 \, m \, s^{-1}$, five seconds after it passed A. Find the distance between A and B.

5. A car accelerates uniformly while travelling on a straight road. The car passes two signposts $360 \, m$ apart. The car takes $15 \, s$ to travel from one signpost to the other. When passing the second signpost, it has speed $28 \, m \, s^{-1}$. Find the speed of the car at the first signpost.

6. A particle is moving along a straight line with constant deceleration. The points X and Y are on the line and $XY = 120 \, m$. At time $t = 0$, the particle passes X and is moving towards Y with speed $18 \, m \, s^{-1}$. At time $t = 10 \, s$, the particle is at Y. Find the velocity of the particle at time $t = 10 \, s$.

7. A cyclist is moving along a straight road from A to B with constant acceleration $0.5 \, m \, s^{-2}$. Her speed at A is $3 \, m \, s^{-1}$ and it takes her 12 seconds to cycle from A to B. Find **a** her speed at B, **b** the distance from A to B.

8. A particle is moving along a straight line with constant acceleration from a point A to a point B, where $AB = 24 \, m$. The particle takes $6 \, s$ to move from A to B and the speed of the particle at B is $5 \, m \, s^{-1}$. Find **a** the speed of the particle at A, **b** the acceleration of the particle.

9. A particle moves in a straight line from a point A to a point B with constant deceleration $1.2 \, m \, s^{-2}$. The particle takes $6 \, s$ to move from A to B. The speed of the particle at B is $2 \, m \, s^{-1}$ and the direction of motion of the particle has not changed. Find **a** the speed of the particle at A, **b** the distance from A to B.

10. A train, travelling on a straight track, is slowing down with constant deceleration $0.6 \, m \, s^{-2}$. The train passes one signal with speed $72 \, km \, h^{-1}$ and a second signal $25 \, s$ later. Find **a** the speed, in $km \, h^{-1}$, of the train as it passes the second signal, **b** the distance between the signals.

11. A particle moves in a straight line from a point A to a point B with a constant deceleration of $4 \, m \, s^{-2}$. At A the particle has speed $32 \, m \, s^{-1}$ and the particle comes to rest at B. Find **a** the time taken for the particle to travel from A to B, **b** the distance between A and B.

12 A skier travelling in a straight line up a hill experiences a constant deceleration. At the bottom of the hill, the skier has a speed of $16\,\text{m s}^{-1}$ and, after moving up the hill for $40\,\text{s}$, he comes to rest. Find **a** the deceleration of the skier, **b** the distance from the bottom of the hill to the point where the skier comes to rest.

13 A particle is moving in a straight line with constant acceleration. The points A, B and C lie on this line. The particle moves from A through B to C. The speed of the particle at A is $2\,\text{m s}^{-1}$ and the speed of the particle at B is $7\,\text{m s}^{-1}$. The particle takes $20\,\text{s}$ to move from A to B.

  a Find the acceleration of the particle.

  The speed of the particle is C is $11\,\text{m s}^{-1}$. Find

  b the time taken for the particle to move from B to C,

  c the distance between A and C.

14 A particle moves in a straight line from A to B with constant acceleration $1.5\,\text{m s}^{-2}$. It then moves, along the same straight line, from B to C with a different acceleration. The speed of the particle at A is $1\,\text{m s}^{-1}$ and the speed of the particle at C is $43\,\text{m s}^{-1}$. The particle takes $12\,\text{s}$ to move from A to B and $10\,\text{s}$ to move from B to C. Find

  a the speed of the particle at B,

  b the acceleration of the particle as it moves from B to C,

  c the distance from A to C.

15 A cyclist travels with constant acceleration $x\,\text{m s}^{-2}$, in a straight line, from rest to $5\,\text{m s}^{-1}$ in $20\,\text{s}$. She then decelerates from $5\,\text{m s}^{-1}$ to rest with constant deceleration $\frac{1}{2}x\,\text{m s}^{-2}$. Find **a** the value of x, **b** the total distance she travelled.

16 A particle is moving with constant acceleration in a straight line. It passes through three points, A, B and C with speeds $20\,\text{m s}^{-1}$, $30\,\text{m s}^{-1}$ and $45\,\text{m s}^{-1}$ respectively. The time taken to move from A to B is t_1 seconds and the time taken to move from B to C is t_2 seconds.

  a Show that $\dfrac{t_1}{t_2} = \dfrac{2}{3}$.

  Given also that the total time taken for the particle to move from A to C is $50\,\text{s}$,

  b find the distance between A and B.

2.2 You can use the formulae $v^2 = u^2 + 2as$, $s = ut + \frac{1}{2}at^2$ and $s = vt - \frac{1}{2}at^2$ for a particle moving in a straight line with constant acceleration.

You can eliminate t from the formulae for constant acceleration.

$t = \dfrac{v - u}{a}$

> Rearrange the formula $v = u + at$ to make t the subject.

$s = \left(\dfrac{u + v}{2}\right)\left(\dfrac{v - u}{a}\right)$

$2as = v^2 - u^2$

> Substitute this expression for t into $s = \left(\dfrac{u + v}{2}\right)t$.

■ $v^2 = u^2 + 2as$

> Multiply out the brackets and rearrange.

You can also eliminate v from the formulae for constant acceleration.

$$s = \left(\frac{u + u + at}{2}\right)t$$

Substitute $v = u + at$ into $s = \left(\frac{u + v}{2}\right)t$.

$$= \left(\frac{2u}{2} + \frac{at}{2}\right)t$$

$$= \left(u + \frac{1}{2}at\right)t$$

Multiply out the brackets and rearrange.

■ $s = ut + \frac{1}{2}at^2$

Finally, you can eliminate u by substituting into this formula:

$$s = (v - at)t + \frac{1}{2}at^2$$

Substitute $u = v - at$ into $s = ut + \frac{1}{2}at^2$.

■ $s = vt - \frac{1}{2}at^2$

■ **You need to remember the five formulae for solving problems about particles moving in a straight lines with constant acceleration.**

  ○ $v = u + at$

  ○ $s = \left(\frac{u + v}{2}\right)t$

  ○ $v^2 = u^2 + 2as$

You need to remember these formulae. They are not given in the formula booklet in your exam.

  ○ $s = ut + \frac{1}{2}at^2$

  ○ $s = vt - \frac{1}{2}at^2$

Example 5

A particle is moving along a straight line from A to B with constant acceleration $5\,\mathrm{m\,s^{-2}}$. The velocity of the particle at A is $3\,\mathrm{m\,s^{-1}}$ in the direction $\overrightarrow{AB}$. The velocity of the particle at B is $18\,\mathrm{m\,s^{-1}}$ in the same direction. Find the distance from A to B.

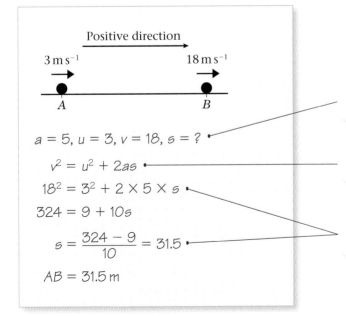

Write down the values you know and the values you need to find. This will help you choose the correct formula.

$a = 5, u = 3, v = 18, s = ?$

$v^2 = u^2 + 2as$

t is not involved so choose the formula that does not have t in it.

$18^2 = 3^2 + 2 \times 5 \times s$

$324 = 9 + 10s$

Substitute in the values you are given and solve the equation for s. This gives the distance you were asked to find.

$s = \dfrac{324 - 9}{10} = 31.5$

$AB = 31.5\,\mathrm{m}$

Example **6**

A car is travelling along a straight horizontal road with a constant acceleration of $0.75\,\mathrm{m\,s^{-2}}$. The car is travelling at $8\,\mathrm{m\,s^{-1}}$ when it passes a pillar box. 12 seconds later it passes a lamp post. Find **a** the distance between the pillar box and the lamp post, **b** the speed with which the car passes the lamp post.

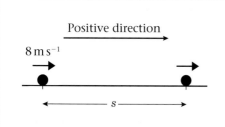

Positive direction

$8\,\mathrm{m\,s^{-1}}$

s

a $a = 0.75,\ u = 8,\ t = 12,\ s = ?$

$$s = ut + \frac{1}{2}at^2$$

$$= 8 \times 12 + \frac{1}{2} \times 0.75 \times 12^2$$

$$= 96 + 54 = 150$$

The distance between the pillar box and the lamp post is 150 m.

b $a = 0.75,\ u = 8,\ t = 12,\ v = ?$

$$v = u + at$$

$$= 8 + 0.75 \times 12$$

$$= 17\,\mathrm{m\,s^{-1}}$$

The speed of the car at the lamp post is $17\,\mathrm{m\,s^{-1}}$.

You are given a, u and t and asked to find s. The final velocity, v, is not given or asked for, so choose the formula without v.

In part **b**, you can use the same values for a, u and t but you are now asked to find v. In this part s is not needed, so choose the formula without s.

You could also solve part **b** using the value of s found in part **a** and the formula $v^2 = u^2 + 2as$.

There is an element of risk in using $s = 150$. This is your answer to part **a** and everyone makes mistakes from time to time. In examinations, it is a good plan to use the data given in a question, rather than your own answer, unless this is impossible or causes a lot of extra work.

Example **7**

A particle moves with constant acceleration $1.5\,\mathrm{m\,s^{-2}}$ in a straight line from a point A to a point B, where $AB = 16\,\mathrm{m}$. At A, the particle has speed $3\,\mathrm{m\,s^{-1}}$. Find the speed of the particle at B.

$a = 1.5,\ u = 3,\ s = 16,\ v = ?$

$v^2 = u^2 + 2as$

$\quad = 3^2 + 2 \times 1.5 \times 16$

$v = \sqrt{57} \approx 7.5498...$

The speed of the particle at B is

$7.55\,\mathrm{m\,s^{-1}}$, to three significant figures.

There is no t here, so choose the formula without t.

$v^2 = 57$ has two possible solutions, $v = \pm\sqrt{57}$. Look at the question to decide whether you need the positive or negative solution (or both). The particle has positive acceleration and positive initial speed, so v must be positive.

It is usually reasonable to give your answer to three significant figures.

Example 8

A particle is moving in a straight horizontal line with constant deceleration $4\,\text{m s}^{-2}$. At time $t = 0$ the particle passes through a point O with speed $13\,\text{m s}^{-1}$ travelling towards a point A where $OA = 20\,\text{m}$. Find **a** the times when the particle passes through A, **b** the velocities of the particle when it passes through A, **c** the values of t when the particle returns to O.

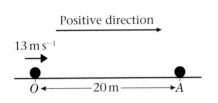

Positive direction

$13\,\text{m s}^{-1}$

$O \xleftarrow{\hspace{2.2cm}} 20\,\text{m} \xrightarrow{\hspace{0.6cm}} A$

The particle is decelerating so the value of a is negative.

a $a = -4$, $u = 13$, $s = 20$, $t = ?$

$s = ut + \dfrac{1}{2}at^2$

$20 = 13t - \dfrac{1}{2} \times 4t^2$

$\quad = 13t - 2t^2$

$2t^2 - 13t + 20 = 0$

$(2t - 5)(t - 4) = 0$

$t = \dfrac{5}{2}, 4$

You are told the values of a, u and s and asked to find t. You are given no information about v and are not asked to find it so you choose the formula without v.

This is a quadratic equation. You can solve it using factorisation, or by using the quadratic formula: $x = \dfrac{-b \pm \sqrt{b^2 - 4ac}}{2a}$

The particle moves through A twice, $2\dfrac{1}{2}$ seconds and 4 seconds after moving through O.

There are two answers. Both are correct. The particle moves from O to A, goes beyond A and then turns round and returns to A.

b $u = 13$, $a = -4$, $t = \dfrac{5}{2}$, $v = ?$

$v = u + at$

$\quad = 13 - 4 \times \dfrac{5}{2}$

$\quad = 3$

There are two values of t and you have to find the velocity of both. The formula $v = u + at$ is the simplest one to use.

This answer is positive, so the particle is moving in the positive direction (away from O).

$u = 13$, $a = -4$, $t = 4$, $v = ?$

$v = u + at$

$\quad = 13 - 4 \times 4$

$\quad = -3$

The value of v is negative when $t = 4$, so the particle is moving in the negative direction (towards O).

When $t = \dfrac{5}{2}$, the particle passes through A with velocity $3\,\text{m s}^{-1}$ in the direction $\overrightarrow{OA}$.

When $t = 4$, the particle passes through A with velocity $3\,\text{m s}^{-1}$ in the direction $\overrightarrow{AO}$.

Remember a velocity has a direction as well as a magnitude. Velocity is a vector quantity. When you are asked for a velocity, your answer must contain a direction as well as a magnitude.

c   The particle returns to O when
$s = 0$.

$s = 0, u = 13, a = -4, t = ?$

$s = ut + \dfrac{1}{2}at^2$

$0 = 13t - 2t^2$

$\quad = t(13 - 2t)$

$t = 0, \dfrac{13}{2}$

The particle returns to O
6.5 seconds after it first passed
through O.

> When the particle returns to O, its displacement (distance) from O is zero.

> The first solution ($t = 0$) represents the starting position of the particle. The other solution ($t = \frac{13}{2}$) tells you when the particle returns to O.

Example 9

A cyclist is moving along a straight road with constant acceleration. She first passes a shop and 10 seconds later, travelling at $8\,\text{m s}^{-1}$, she passes a street sign. The distance between the shop and the street sign is 60 m. Find **a** the acceleration of the cyclist, **b** the speed with which she passed the shop.

a $t = 10, v = 8, s = 60, a = ?$

$\quad s = vt - \dfrac{1}{2}at^2$

$\quad 60 = 8 \times 10 - \dfrac{1}{2} \times a \times 100$

$\quad 60 = 80 - 50a$

$\quad 50a = 80 - 60 = 20$

$\quad a = \dfrac{20}{50} = 0.4$

The acceleration of the cyclist is
$0.4\,\text{m s}^{-2}$.

> There is no u, so you choose the formula without u.

> Substitute into the formula and solve the equation for a.

b $v = 8, t = 10, a = 0.4, u = ?$

$\quad v = u + at$

$\quad 8 = u + 10 \times 0.4$

$\quad u = 8 - 10 \times 0.4 = 4$

The cyclist passes the shop with
speed $4\,\text{m s}^{-1}$.

> $v = u + at$ has been chosen as it is a simple formula. You could also use $s = \left(\dfrac{u + v}{2}\right)t$ which would avoid using your answer to part **a**.

Example 10

A particle P is moving on the x-axis with constant deceleration 2.5 m s^{-2}. At time $t = 0$, the particle P passes through the origin O, moving in the positive direction of x with speed 15 m s^{-1}. Find **a** the time between the instant when P first passes through O and the instant when it returns to O, **b** the total distance travelled by P during this time.

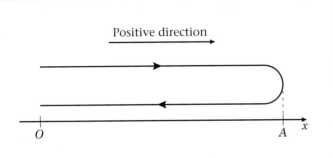

Positive direction

O A x

Before you start, draw a sketch so you can see what is happening. The particle moves through O with a positive velocity. As it is decelerating it slows down and will eventually have zero velocity at a point A, which you don't yet know. As the particle is still decelerating, its velocity becomes negative, so the particle changes direction and returns to O.

a $a = -2.5, u = 15, s = 0, t = ?$

$s = ut + \frac{1}{2}at^2$

$0 = 15t - \frac{1}{2} \times 2.5 \times t^2$

$0 = 60t - 5t^2$

$ = 5t(12 - t)$

$t = 0, t = 12$

The particle P returns to O after 12 s.

When the particle returns to O, its displacement (distance) from O is zero.

Multiply by 4 to get whole-number coefficients.

b $a = -2.5, u = 15, v = 0, s = ?$

$v^2 = u^2 + 2as$

$0^2 = 15^2 - 2 \times 2.5 \times s$

$5s = 15^2 = 225$

$s = \frac{225}{5} = 45$

The distance $OA = 45 \text{ m}$.

The total distance travelled by P is $2 \times 45 \text{ m} = 90 \text{ m}$.

At the furthest point from O, labelled A in the diagram, the particle changes direction. At that point, for an instant, the particle has zero velocity.

In the 12 s the particle has been moving it has travelled to A and back. The total distance travelled is twice the distance OA.

Exercise 2B

1   A particle is moving in a straight line with constant acceleration 2.5 m s^{-2}. It passes a point A with speed 3 m s^{-1} and later passes through a point B, where $AB = 8 \text{ m}$. Find the speed of the particle as it passes through B.

2   A car is accelerating at a constant rate along a straight horizontal road. Travelling at 8 m s^{-1}, it passes a pillar box and 6 s later it passes a sign. The distance between the pillar box and the sign is 60 m. Find the acceleration of the car.

3 A cyclist travelling at $12\,\text{m s}^{-1}$ applies her brakes and comes to rest after travelling 36 m in a straight line. Assuming that the brakes cause the cyclist to decelerate uniformly, find the deceleration.

4 A particle moves along a straight line from P to Q with constant acceleration $1.5\,\text{m s}^{-2}$. The particle takes 4 s to pass from P to Q and $PQ = 22$ m. Find the speed of the particle at Q.

5 A particle is moving along a straight line OA with constant acceleration $2\,\text{m s}^{-2}$. At O the particle is moving towards A with speed $5.5\,\text{m s}^{-1}$. The distance OA is 20 m. Find the time the particle takes to move from O to A.

6 A train is moving along a straight horizontal track with constant acceleration. The train passes a signal at $54\,\text{km h}^{-1}$ and a second signal at $72\,\text{km h}^{-1}$. The distance between the two signals is 500 m. Find, in m s^{-2}, the acceleration of the train.

7 A particle moves along a straight line, with constant acceleration, from a point A to a point B where $AB = 48$ m. At A the particle has speed $4\,\text{m s}^{-1}$ and at B it has speed $16\,\text{m s}^{-1}$. Find **a** the acceleration of the particle, **b** the time the particle takes to move from A to B.

8 A particle moves along a straight line with constant acceleration $3\,\text{m s}^{-2}$. The particle moves 38 m in 4 s. Find **a** the initial speed of the particle, **b** the final speed of the particle.

9 The driver of a car is travelling at $18\,\text{m s}^{-1}$ along a straight road when she sees an obstruction ahead. She applies the brakes and the brakes cause the car to slow down to rest with a constant deceleration of $3\,\text{m s}^{-2}$. Find **a** the distance travelled as the car decelerates, **b** the time it takes for the car to decelerate from $18\,\text{m s}^{-1}$ to rest.

10 A stone is sliding across a frozen lake in a straight line. The initial speed of the stone is $12\,\text{m s}^{-1}$. The friction between the stone and the ice causes the stone to slow down at a constant rate of $0.8\,\text{m s}^{-2}$. Find **a** the distance moved by the stone before coming to rest, **b** the speed of the stone at the instant when it has travelled half of this distance.

11 A particle is moving along a straight line OA with constant acceleration $2.5\,\text{m s}^{-2}$. At time $t = 0$, the particle passes through O with speed $8\,\text{m s}^{-1}$ and is moving in the direction OA. The distance OA is 40 m. Find **a** the time taken for the particle to move from O to A, **b** the speed of the particle at A. Give your answers to one decimal place.

12 A particle travels with uniform deceleration $2\,\text{m s}^{-2}$ in a horizontal line. The points A and B lie on the line and $AB = 32$ m. At time $t = 0$, the particle passes through A with velocity $12\,\text{m s}^{-1}$ in the direction $\overrightarrow{AB}$. Find **a** the values of t when the particle is at B, **b** the velocity of the particle for each of these values of t.

13 A particle is moving along the x-axis with constant deceleration $5\,\text{m s}^{-2}$. At time $t = 0$, the particle passes through the origin O with velocity $12\,\text{m s}^{-1}$ in the positive direction. At time t seconds the particle passes through the point A with x-coordinate 8. Find **a** the values of t, **b** the velocity of the particle as it passes through the point with x-coordinate -8.

14 A particle P is moving on the x-axis with constant deceleration $4\,\mathrm{m\,s^{-2}}$. At time $t = 0$, P passes through the origin O with velocity $14\,\mathrm{m\,s^{-1}}$ in the positive direction. The point A lies on the axis and $OA = 22.5\,\mathrm{m}$. Find **a** the difference between the times when P passes through A, **b** the total distance travelled by P during the interval between these times.

15 A car is travelling along a straight horizontal road with constant acceleration. The car passes over three consecutive points A, B and C where $AB = 100\,\mathrm{m}$ and $BC = 300\,\mathrm{m}$. The speed of the car at B is $14\,\mathrm{m\,s^{-1}}$ and the speed of the car at C is $20\,\mathrm{m\,s^{-1}}$. Find **a** the acceleration of the car, **b** the time take for the car to travel from A to C.

16 Two particles P and Q are moving along the same straight horizontal line with constant accelerations $2\,\mathrm{m\,s^{-2}}$ and $3.6\,\mathrm{m\,s^{-2}}$ respectively. At time $t = 0$, P passes through a point A with speed $4\,\mathrm{m\,s^{-1}}$. One second later Q passes through A with speed $3\,\mathrm{m\,s^{-1}}$, moving in the same direction as P.

   a Write down expressions for the displacements of P and Q from A, in terms of t, where t seconds is the time after P has passed through A.

   b Find the value of t where the particles meet.

   c Find the distance of A from the point where the particles meet.

2.3 **You can use the formulae for constant acceleration to model an object moving vertically in a straight line under gravity.**

■ The force of gravity causes all objects to accelerate towards the earth. If you ignore the effects of air resistance, this acceleration is constant. It does not depend on the mass of the object.

This means that in a vacuum an apple and a feather would both accelerate downwards at the same rate.

■ On earth, the acceleration due to gravity is represented by the letter g and is approximately $9.8\,\mathrm{m\,s^{-2}}$. The actual value of the acceleration can vary by very small amounts in different places due to the changing radius of the Earth and height above sea level.

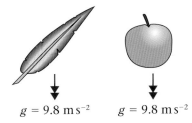

$g = 9.8\ \mathrm{m\,s^{-2}}$ $g = 9.8\ \mathrm{m\,s^{-2}}$

Objects falling in a vacuum accelerate at the same rate regardless of their mass or size.

> In M1 you will always use $g = 9.8\,\mathrm{m\,s^{-2}}$. This is an approximation to two significant figures. If you use this value in your working you should give your answer to the same degree of accuracy.

■ An object moving vertically in a straight line can be modelled as a particle with a constant downward acceleration of $g = 9.8\,\mathrm{m\,s^{-2}}$

■ When solving problems about vertical motion you can choose the positive direction to be either upwards or downwards. Acceleration due to gravity is always downwards, so if the positive direction is upwards then $a = -9.8\,\mathrm{m\,s^{-2}}$.

■ The total time that an object is in motion from the time it is projected (thrown) upwards to the time it hits the ground is called the **time of flight**. The initial speed is sometimes called the **speed of projection**.

Example 11

A ball B is projected vertically upwards from a point O with speed $12\,\text{m s}^{-1}$. Find **a** the greatest height above O reached by B, **b** the total time before B returns to O.

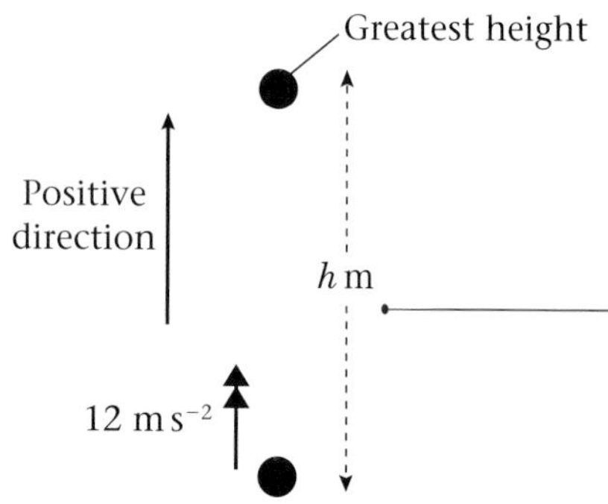

Greatest height

Positive direction

h m

$12\,\text{m s}^{-2}$

You first decide which direction you will take as positive. As the ball is projected upwards, you take the upwards direction as positive.

Writing the unknown is h metres helps you to remember you are finding a height!

a $u = 12$

$v = 0$

$a = -9.8$

$s = h$

$v^2 = u^2 + 2as$

$0^2 = 12^2 - 2 \times 9.8 \times h$

$h = \dfrac{12^2}{2 \times 9.8} = \dfrac{144}{19.6} = 7.346\ldots$

The greatest height above O reached by B is $7.4\,\text{m}$, to two significant figures.

At the highest point of its path, the ball is turning round. For an instant, its speed is zero.

The positive direction is upwards, and gravity acts downwards, so a is negative.

Write your answer correct to two significant figures.

b $s = 0$

$u = 12$

$a = -9.8$

$t = ?$

$s = ut + \dfrac{1}{2}at^2$

$0 = 12t - \dfrac{1}{2} \times 9.8 \times t = t(12 - 4.9t)$

$t = \dfrac{12}{4.9} = 2.448\ldots$

The time taken for B to return to O is $2.4\,\text{s}$, to two significant figures.

When the ball returns to its original position O, its displacement from O is zero.

This equation has two answers but one of them is $t = 0$. This represents the start of the ball's motion, so you want the other answer.

You must work to at least three significant figures and correct your answer to two significant figures.

Example 12

A book falls off the top shelf of a bookcase. The shelf is 1.4 m above a wooden floor. Find **a** the time the book takes to reach the floor, **b** the speed with which the book strikes the floor.

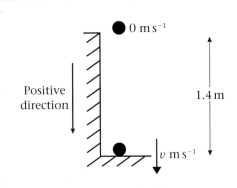

Model the book as a particle moving in a straight line with a constant acceleration of magnitude $9.8\,\text{m s}^{-2}$.

As the book is moving downwards throughout its motion, it is sensible to take the downwards direction as positive.

a $s = 1.4$

$a = +9.8$

$u = 0$

$t = ?$

$s = ut + \frac{1}{2}at^2$

$1.4 = 0 + \frac{1}{2} \times 9.8 \times t^2$

$t^2 = \dfrac{1.4}{4.9} = 0.2857\ldots$

$t = \sqrt{0.2857\ldots} = 0.5345\ldots$

The time taken for the book to reach the floor is 0.53 s, to two significant figures.

You have taken the downwards direction as positive and gravity acts downwards. Here the acceleration is positive.

Assume the book has an initial speed of zero.

Choose the formula without v.

Solve the equation for t^2 and use your calculator to find the positive square root.

b $s = 1.4$

$a = 9.8$

$u = 0$

$v = ?$

$v^2 = u^2 + 2as$

$\quad = 0^2 + 2 \times 9.8 \times 1.4 = 27.44$

$v = \sqrt{27.44} = 5.238\ldots \approx 5.2$

The book hits the floor with speed $5.2\,\text{m s}^{-1}$, to two significant figures.

Remember to give the answer to two significant figures.

Choose the formula without t.

Remember to show your working to at least three significant figures. You can use unrounded values in your calculations by using the Ans button on your calculator.

Example 13

A ball is projected vertically upwards, from a point X which is 7 m above the ground, with speed 21 m s^{-1}. Find **a** the greatest height above the ground reached by the ball, **b** the time of flight of the ball.

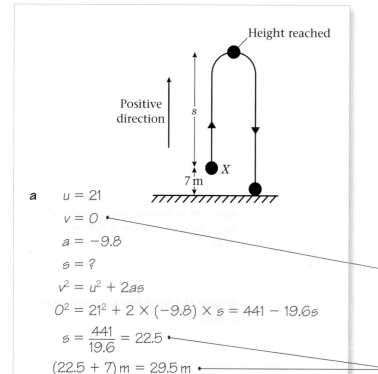

In this sketch the upward and downwards motion have been sketched side by side. In reality they would be on top of one another, but drawing them separately makes it easier to see what is going on.

You model the ball as a particle moving in a straight line with a constant acceleration of magnitude 9.8 m s^{-2}.

a
$u = 21$
$v = 0$
$a = -9.8$
$s = ?$
$v^2 = u^2 + 2as$
$0^2 = 21^2 + 2 \times (-9.8) \times s = 441 - 19.6s$
$s = \dfrac{441}{19.6} = 22.5$
$(22.5 + 7)\,\text{m} = 29.5\,\text{m}$

The greatest height reached by the ball above the ground is 30 m, to two significant figures.

At its highest point, the ball is turning round. For an instant, it is neither going up or down, so its speed is zero.

22.5 m is the distance the ball has moved above X but X is 7 m above the ground. You must add on another 7 m to get the greatest height above the ground reached by the ball.

b
$s = -7$
$u = 21$
$a = -9.8$
$t = ?$
$s = ut + \dfrac{1}{2}at^2$
$-7 = 21t - 4.9t^2$
$4.9t^2 - 21t - 7 = 0$
$t = \dfrac{-b \pm \sqrt{(b^2 - 4ac)}}{2a}$
$= \dfrac{-(-21) \pm \sqrt{((-21)^2 - 4 \times 4.9 \times (-7))}}{2 \times 4.9}$
$= \dfrac{21 \pm \sqrt{578.2}}{9.8} \approx \dfrac{21 \pm 24.046}{9.8}$
$\approx 4.5965, -0.3108$

The time of flight of the ball is 4.6 s, to two significant figures.

The time of flight is the total time that the ball is in motion from the time that it is projected to the time that it stops moving. Here the ball will stop when it hits the ground. The point where the ball hits the ground is 7 m **below** the point from which it was projected so $s = -7$.

Rearrange the equation and use the quadratic formula.

The negative answer represents a time before the particle was projected, so you need the positive answer. Remember to give your answer to two significant figures.

Example 14

A particle is projected vertically upwards from a point O with speed $u\,\mathrm{m\,s^{-1}}$. The greatest height reached by the particle is 62.5 m above O. Find **a** the value of u, **b** the total time for which the particle is 50 m or more above O.

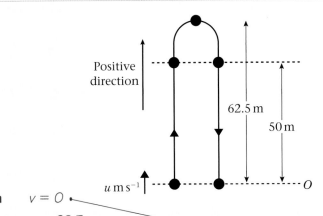

The particle will pass through the point 50 m above O twice. Once on the way up and once on the way down.

a $v = 0$

$s = 62.5$

$a = -9.8$

$u = ?$

There is no t, so you choose the formula without t.

$v^2 = u^2 + 2as$

$0^2 = u^2 - 2 \times 9.8 \times 62.5$

$u^2 = 1225$

$u = \sqrt{1225} = 35$

In this part, you obtain an exact answer, so there is no need for approximation.

b $s = 50$

$u = 35$

$a = -9.8$

$t = ?$

Two values of t need to be found: one on the way up and one on the way down.

$s = ut + \frac{1}{2}at^2$

$50 = 35t - 4.9t^2$

$4.9t^2 - 35t + 50 = 0$

Write this equation in the form $ax^2 + bx + c = 0$ and use the quadratic formula.

$t = \dfrac{-b \pm \sqrt{(b^2 - 4ac)}}{2a}$

$= \dfrac{35 \pm \sqrt{(35^2 - 4 \times 4.9 \times 50)}}{9.8}$

$= \dfrac{35 \pm \sqrt{245}}{9.8} = \dfrac{35 \pm 15.6525}{9.8}$

$= 5.1686\ldots, 1.9742\ldots$

$(5.1686\ldots) - (1.9742\ldots) \approx 3.194$

Between these two times the particle is always more than 50 m above O. You find the total time for which the particle is 50 m or more above O by finding the difference of these two values.

The total time for which the particle is 50 m or more above O is 3.2 m, to two significant figures.

Example **15**

A ball A falls vertically from rest from the top of a tower 63 m high. At the same time as A begins to fall, another ball B is projected vertically upwards from the bottom of the tower with speed $21\,\text{m s}^{-1}$. The balls collide. Find the distance of the point where the balls collide from the bottom of the tower.

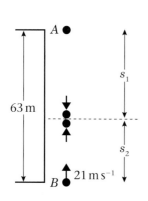

For A, the motion is downwards

$u = 0$

$a = 9.8$

$s = ut + \dfrac{1}{2}at^2$

$s_1 = 4.9t^2$

For B, the motion is upwards

$u = 21$

$a = -9.8$

$s = ut + \dfrac{1}{2}at^2$

$s_2 = 21t - 4.9t^2$

The height of the tower is 63 m.

$s_1 + s_2 = 63$

$4.9t^2 + (21t - 4.9t^2) = 63$

$21t = 63$

$t = 3$

$s_2 = 21t - 4.9t^2$

$= 21 \times 3 - 4.9 \times 3^2 = 18.9$

The balls collide 19 m from the bottom of the tower, to two significant figures.

You must take special care with problems where objects are moving in different directions. Here A is moving downwards and you will take the acceleration due to gravity as positive. However B is moving upwards so for B the acceleration due to gravity is negative.

You cannot find s_1 at this stage. You have to express it in terms of t.

As B is moving upwards, the acceleration due to gravity is negative.

You now have expressions for s_1 and s_2 in terms of t.

Adding together the two distances gives the height of the tower. You can write this as an equation in t.

The $4.9t^2$ and $-4.9t^2$ cancel.

You have found t but you were asked for the distance from the bottom of the tower. Substitute your value for t into your equation for s_2.

Exercise 2C

1 A ball is projected vertically upwards from a point O with speed $14\,\mathrm{m\,s^{-1}}$. Find the greatest height above O reached by the ball.

2 A well is $50\,\mathrm{m}$ deep. A stone is released from rest at the top of the well. Find how long the stone takes to reach the bottom of the well.

3 A book falls from the top shelf of a bookcase. It takes $0.6\,\mathrm{s}$ to reach the floor. Find how far it is from the top shelf to the floor.

4 A particle is projected vertically upwards with speed $20\,\mathrm{m\,s^{-1}}$ from a point on the ground. Find the time of flight of the particle.

5 A ball is thrown vertically downward from the top of a tower with speed $18\,\mathrm{m\,s^{-1}}$. It reaches the ground in $1.6\,\mathrm{s}$. Find the height of the tower.

6 A pebble is catapulted vertically upwards with speed $24\,\mathrm{m\,s^{-1}}$. Find **a** the greatest height above the point of projection reached by the pebble, **b** the time taken to reach this height.

7 A ball is projected upwards from a point which is $4\,\mathrm{m}$ above the ground with speed $18\,\mathrm{m\,s^{-1}}$. Find **a** the speed of the ball when it is $15\,\mathrm{m}$ above its point of projection, **b** the speed with which the ball hits the ground.

8 A particle P is projected vertically downwards from a point $80\,\mathrm{m}$ above the ground with speed $4\,\mathrm{m\,s^{-1}}$. Find **a** the speed with which P hits the ground, **b** the time P takes to reach the ground.

9 A particle P is projected vertically upwards from a point X. Five seconds later P is moving downwards with speed $10\,\mathrm{m\,s^{-1}}$. Find **a** the speed of projection of P, **b** the greatest height above X attained by P during its motion.

10 A ball is thrown vertically upwards with speed $21\,\mathrm{m\,s^{-1}}$. It hits the ground $4.5\,\mathrm{s}$ later. Find the height above the ground from which the ball was thrown.

11 A stone is thrown vertically upward from a point which is $3\,\mathrm{m}$ above the ground, with speed $16\,\mathrm{m\,s^{-1}}$. Find **a** the time of flight of the stone, **b** the total distance travelled by the stone.

12 A particle is projected vertically upwards with speed $24.5\,\mathrm{m\,s^{-1}}$. Find the total time for which it is $21\,\mathrm{m}$ or more above its point of projection.

13 A particle is projected vertically upwards from a point O with speed $u\,\mathrm{m\,s^{-1}}$. Two seconds later it is still moving upwards and its speed is $\frac{1}{3}u\,\mathrm{m\,s^{-1}}$. Find **a** the value of u, **b** the time from the instant that the particle leaves O to the instant that it returns to O.

14 A ball A is thrown vertically downwards with speed $5\,\mathrm{m\,s^{-1}}$ from the top of a tower block $46\,\mathrm{m}$ above the ground. At the same time as A is thrown downwards, another ball B is thrown vertically upwards from the ground with speed $18\,\mathrm{m\,s^{-1}}$. The balls collide. Find the distance of the point where A and B collide from the point where A was thrown.

15 A ball is released from rest at a point which is 10 m above a wooden floor. Each time the ball strikes the floor, it rebounds with three-quarters of the speed with which it strikes the floor. Find the greatest height above the floor reached by the ball **a** the first time it rebounds from the floor, **b** the second time it rebounds from the floor.

16 A particle P is projected vertically upwards from a point O with speed $12\,\text{m s}^{-1}$. One second after P has been projected from O, another particle Q is projected vertically upwards from O with speed $20\,\text{m s}^{-1}$. Find **a** the time between the instant that P is projected from O and the instant when P and Q collide, **b** the distance of the point where P and Q collide from O.

2.4 You can represent the motion of an object on a speed–time graph or a distance–time graph

■ In a speed–time graph speed is always plotted on the vertical axis and time is always plotted on the horizontal axis. This speed–time graph represents the motion of a particle accelerating from speed u at time 0 to speed v at time t.

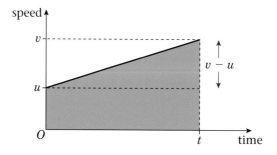

Gradient of line $= \dfrac{\text{change of velocity}}{\text{time}}$

$$= \frac{v - u}{t} = a$$

So the gradient of the speed–time graph is the acceleration of the particle. If the line is straight the acceleration is constant.

Using the formula for the area of a trapezium:

Shaded area $= \left(\dfrac{u + v}{2}\right)t$

$$= s$$

So the area under the speed–time graph is the distance travelled by the particle.

■ **The gradient of a speed–time graph is the acceleration.**

■ **The area under a speed–time graph is the distance travelled.**

You can also draw acceleration–time graphs and distance–time graphs for the motion of a particle. Time is always plotted on the horizontal axis.

If a particle is moving with constant speed its distance–time graph will be a straight line. If it is accelerating or decelerating then its distance–time graph will be a curve.

Example 16

A car accelerates uniformly from rest for 20 seconds. It travels at a constant speed for the next 40 seconds, then decelerates uniformly for 20 seconds until it is stationary. Sketch **a** an acceleration–time graph, **b** a distance–time graph for the motion of the car.

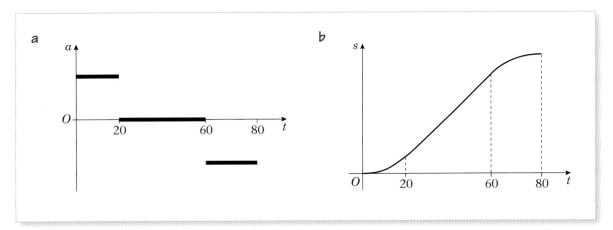

Example 17

The figure shows a speed–time graph illustrating the motion of a cyclist moving along a straight road for a period of 12 s. For the first 8 s, she moves at a constant speed of $6 \, \text{m s}^{-1}$. She then decelerates at a constant rate, stopping after a further 4 s.

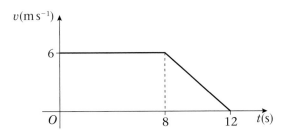

Find **a** the distance travelled by the cyclist during this 12 s period, **b** the rate at which the cyclist decelerates.

Model the cyclist as a particle moving in a straight line.

a The distance travelled is given by

$$s = \frac{1}{2}(a + b)h$$

$$= \frac{1}{2}(8 + 12) \times 6$$

$$= 10 \times 6 = 60$$

The distance travelled by the cyclist is 60 m.

The distance travelled is represented by the area of the trapezium with these sides.

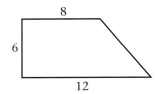

You can use the formula for the area of a trapezium to calculate this area.

b The acceleration is the gradient of the slope.

$$a = \frac{-6}{4} = -1.5$$

The deceleration is $1.5 \, \mathrm{m \, s^{-2}}$.

The gradient is given by

$$\frac{\text{the difference in the } v \text{ coordinates}}{\text{the difference in the } t \text{ coordinates}}$$

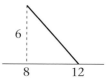

Here, the value of v **decreases** by 6 as t **increases** by 4.

Example 18

A car is waiting at traffic lights. When the lights turn green, the car accelerates uniformly from rest to a speed of $10 \, \mathrm{m \, s^{-1}}$ in $20 \, \mathrm{s}$. This speed is then maintained until the car passes a road sign $50 \, \mathrm{s}$ after leaving the traffic lights.

a Sketch a speed–time graph to illustrate the motion of the car.

b Find the distance between the traffic lights and the road sign.

Model the car as a particle moving in a straight line.

a

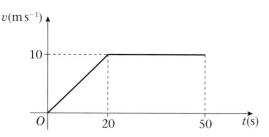

When you are asked to sketch a graph you should use ordinary paper and not graph paper. You should use a ruler but you do not have to draw lengths accurately to scale. You should label the axes and indicate any relevant information given in the question.

b The distance travelled is given by

$$s = \frac{1}{2}(a + b)h$$

$$= \frac{1}{2}(30 + 50) \times 10$$

$$= 40 \times 10 = 400$$

The distance between the traffic lights and the road sign is $400 \, \mathrm{m}$.

The dimensions of the trapezium are:

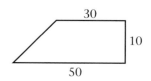

Example 19

A particle moves along a straight line. The particle accelerates uniformly from rest to a speed of $8 \, m \, s^{-1}$ in T seconds. The particle then travels at a constant speed of $8 \, m \, s^{-1}$ for $5T$ seconds. The particle then decelerates uniformly to rest in a further $40 \, s$.

a Sketch a speed–time graph to illustrate the motion of the particle.

Given that the total distance travelled by the particle is $600 \, m$, **b** find the value of T, **c** sketch an acceleration–time graph illustrating the motion of the particle.

a

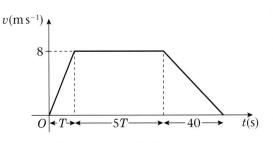

b The area of the trapezium is:

$$s = \frac{1}{2}(a + b)h$$

$$= \frac{1}{2}(5T + 6T + 40) \times 8$$

$$= 4(11T + 40)$$

The distance moved is $600 \, m$.

$$4(11T + 40) = 600$$

$$44T + 160 = 600$$

$$T = \frac{600 - 160}{44} = 10$$

> The length of the shorter of the two parallel sides is $5T$. The length of the longer side is $T + 5T + 40 = 6T + 40$.

> The distance moved is equal to the area of the trapezium. Write an equation and solve it to find T.

c The acceleration in the first $10 \, s$ is given by

$$a = \frac{8}{10} = 0.8.$$

The acceleration in the last $40 \, s$ is given by

$$a = \frac{-8}{40} = -0.2$$

> For the first ten seconds the v-coordinate **increases** by 8 as the t-coordinate increases by 10. This gives a positive answer.

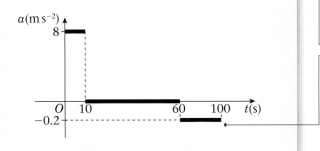

> In the last forty seconds the v-coordinate **decreases** by 8 as the t-coordinate increases by 40. This gives a negative answer.

Example 20

A car C is moving along a straight road with constant speed $17.5 \, \text{m s}^{-1}$. At time $t = 0$, C passes a lay-by. At time $t = 0$, a second car D leaves the lay-by. Car D accelerates from rest to a speed of $20 \, \text{m s}^{-1}$ in 15 s and then maintains a constant speed of $20 \, \text{m s}^{-1}$. Car D passes car C at a road sign.

a On the same diagram, sketch speed–time graphs to illustrate the motion of the two cars.

b Find the distance between the lay-by and the road sign.

a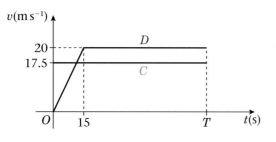

You should label the lines so that it is clear. which represents the motion of C and which represents the motion of D.

b Let D pass C at time $t = T$.

It is difficult to find the distance travelled directly. You can find the time the cars pass first. It does not matter what letter you choose for the time. T has been used here.

The distance travelled by C is given by

$$s = 17.5T$$

As C is travelling at a constant speed you use the formula distance = speed × time.

The distance travelled by D is given by

$$s = \frac{1}{2}(a + b)h$$

$$= \frac{1}{2}(T - 15 + T) \times 20$$

$$= 10(2T - 15)$$

The longer of the parallel sides of the trapezium is T. The shorter of the parallel sides is $(T - 15)$.

The distances travelled by C and D are the same.

$$10(2T - 15) = 17.5T$$

$$20T - 150 = 17.5T$$

$$2.5T = 150$$

$$T = \frac{150}{2.5} = 60$$

As the cars were at the lay-by at the same time and the road sign at the same time, the distance travelled by both of them is the same. You equate the distances to get an equation in T and solve it.

$$s = 17.5T = 17.5 \times 60 = 1050$$

The distance from the lay-by to the road sign is 1050 m.

To find the distance travelled you can substitute into the expression for the distance travelled by either C or by D.

Example 21

A particle is moving along a horizontal axis Ox. At time $t = 0$, the particle is at rest at O. The particle then accelerates at a constant rate, reaching a speed of $u\,\mathrm{m\,s^{-1}}$ in 16 s. The particle maintains the speed of $u\,\mathrm{m\,s^{-1}}$ for a further 32 s. After 48 s, the particle is at a point A, where $OA = 320$ m.

a Sketch a speed–time graph to illustrate the motion of the particle.

b Find the value of u.

c Sketch a distance–time graph for the particle.

a

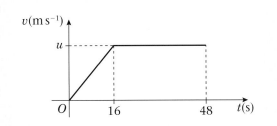

b $\frac{1}{2}(32 + 48) \times u = 320$

$40u = 320$

$u = 8$

The area of the trapezium is 320. You use this to obtain an equation in u, which you solve.

c The gradient of the line in the first 16 s is the acceleration and is given by

$$a = \frac{u}{16} = \frac{8}{16} = \frac{1}{2}$$

Use your answer from part **b** and the property that the gradient of a speed–time graph is the acceleration to find the value of the acceleration.

$$s = ut + \frac{1}{2}at^2$$

The initial speed is zero and the acceleration is $\frac{1}{2}$.

$$= \frac{1}{4}t^2$$

When $t = 16$, $s = \frac{1}{4} \times 16^2 = 64$.

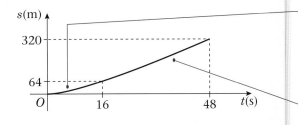

For the first 16 s the distance–time graph is a curve. It is part of the parabola with equation $s = \frac{1}{4}t^2$.

For the following 32 s, the particle moves with constant speed and the graph is a straight line.

Exercise 2D

1

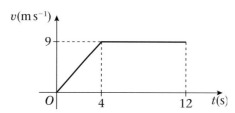

The diagram shows the speed–time graph of the motion of an athlete running along a straight track. For the first 4 s, he accelerates uniformly from rest to a speed of 9 m s⁻¹. This speed is then maintained for a further 8 s. Find

a the rate at which the athlete accelerates,

b the total distance travelled by the athlete in 12 s.

2 A car is moving along a straight road. When $t = 0$ s, the car passes a point A with speed 10 m s⁻¹ and this speed is maintained until $t = 30$ s. The driver then applies the brakes and the car decelerates uniformly, coming to rest at the point B when $t = 42$ s.

a Sketch a speed–time graph to illustrate the motion of the car.

b Find the distance from A to B.

3

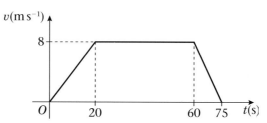

The diagram shows the speed–time graph of the motion of a cyclist riding along a straight road. She accelerates uniformly from rest to 8 m s⁻¹ in 20 s. She then travels at a constant speed of 8 m s⁻¹ for 40 s. She then decelerates uniformly to rest in 15 s. Find

a the acceleration of the cyclist in the first 20 s of motion,

b the deceleration of the cyclist in the last 15 s of motion,

c the total distance travelled in 75 s.

4 A car accelerates at a constant rate, starting from rest at a point A and reaching a speed of 45 km h⁻¹ in 20 s. This speed is then maintained and the car passes a point B 3 minutes after leaving A.

a Sketch a speed–time graph to illustrate the motion of the car.

b Find the distance from A to B.

5 A motorcyclist starts from rest at a point S on a straight race track. He moves with constant acceleration for 15 s, reaching a speed of 30 m s⁻¹. He then travels at a constant speed of 30 m s⁻¹ for T seconds. Finally he decelerates at a constant rate coming to rest at a point F, 25 s after he begins to decelerate.

a Sketch a speed–time graph to illustrate the motion.

Given that the distance between S and F is 2.4 km,

b calculate the time the motorcyclist takes to travel from S to F.

6

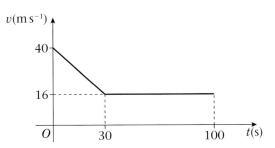

A train is travelling along a straight track. To obey a speed restriction, the brakes of the train are applied for 30 s reducing the speed of the train from $40\,\mathrm{m\,s^{-1}}$ to $16\,\mathrm{m\,s^{-1}}$. The train then continues at a constant speed of $16\,\mathrm{m\,s^{-1}}$ for a further 70 s. The diagram shows a speed–time graph illustrating the motion of the train for the total period of 100 s. Find

a the retardation of the train in the first 30 s.

b the total distance travelled by the train in 100 s.

7 A train starts from a station X and moves with constant acceleration $0.6\,\mathrm{m\,s^{-2}}$ for 20 s. The speed it has reached after 20 s is then maintained for T seconds. The train then decelerates from this speed to rest in a further 40 s, stopping at a station Y.

a Sketch a speed–time graph to illustrate the motion of the train.

Given that the distance between the stations is 4.2 km, find

b the value of T,

c the distance travelled by the train while it is moving with constant speed.

8 A particle moves along a straight line. The particle accelerates from rest to a speed of $10\,\mathrm{m\,s^{-1}}$ in 15 s. The particle then moves at a constant speed of $10\,\mathrm{m\,s^{-1}}$ for a period of time. The particle then decelerates uniformly to rest. The period of time for which the particle is travelling at a constant speed is 4 times the period of time for which it is decelerating.

a Sketch a speed–time graph to illustrate the motion of the particle.

Given that the total distance travelled by the particle is 480 m,

b find the total time for which the particle is moving,

c sketch an acceleration–time graph illustrating the motion of the particle.

9

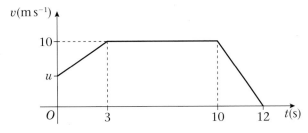

A particle moves 100 m in a straight line. The diagram is a sketch of a speed–time graph of the motion of the particle. The particle starts with speed $u\,\mathrm{m\,s^{-1}}$ and accelerates to a speed $10\,\mathrm{m\,s^{-1}}$ in 3 s. The speed of $10\,\mathrm{m\,s^{-1}}$ is maintained for 7 s and then the particle decelerates to rest in a further 2 s. Find

a the value of u,

b the acceleration of the particle in the first 3 s of motion.

10

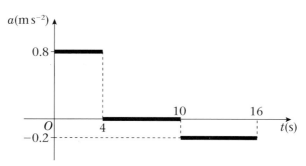

The diagram is an acceleration–time graph to show the motion of a particle. At time $t = 0$ s, the particle is at rest. Sketch a speed–time graph for the motion of the particle.

11 A motorcyclist M leaves a road junction at time $t = 0$ s. She accelerates at a rate of 3 m s^{-2} for 8 s and then maintains the speed she has reached. A car C leaves the same road junction as M at time $t = 0$ s. The car accelerates from rest to 30 m s^{-1} in 20 s and then maintains the speed of 30 m s^{-1}. C passes M as they both pass a pedestrian.

a On the same diagram, sketch speed–time graphs to illustrate the motion of M and C.

b Find the distance of the pedestrian from the road junction.

12 A particle is moving on an axis Ox. From time $t = 0$ s to time $t = 32$ s, the particle is travelling with constant speed 15 m s^{-1}. The particle then decelerates from 15 m s^{-1} to rest in T seconds.

a Sketch a speed–time graph to illustrate the motion of the particle.

The total distance travelled by the particle is 570 m.

b Find the value of T.

c Sketch a distance–time graph illustrating the motion of the particle.

Mixed exercise 2E

1 A car travelling along a straight road at 14 m s^{-1} is approaching traffic lights. The driver applies the brakes and the car comes to rest with constant deceleration. The distance from the point where the brakes are applied to the point where the car comes to rest is 49 m. Find the deceleration of the car.

2 A ball is thrown vertically downward from the top of a tower with speed 6 m s^{-1}. The ball strikes the ground with speed 25 m s^{-1}. Find the time the ball takes to move from the top of the tower to the ground.

3

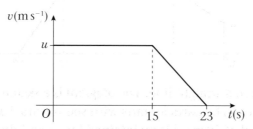

The diagram is a speed–time graph representing the motion of a cyclist along a straight road. At time $t = 0$ s, the cyclist is moving with speed u m s^{-1}. The speed is maintained until time $t = 15$ s, when she slows down with constant deceleration, coming to rest when $t = 23$ s. The total distance she travels in 23 s is 152 m. Find the value of u.

4 A stone is projected vertically upwards with speed $21 \, \text{m s}^{-1}$. Find

    a the greatest height above the point of projection reached by the stone,

    b the time between the instant that the stone is projected and the instant that it reaches its greatest height.

5 A train is travelling with constant acceleration along a straight track. At time $t = 0 \, \text{s}$, the train passes a point O travelling with speed $18 \, \text{m s}^{-1}$. At time $t = 12 \, \text{s}$, the train passes a point P travelling with speed $24 \, \text{m s}^{-1}$. At time $t = 20 \, \text{s}$, the train passes a point Q. Find

    a the speed of the train at Q,

    b the distance from P to Q.

6 A car travelling on a straight road slows down with constant deceleration. The car passes a road sign with speed $40 \, \text{km h}^{-1}$ and a post box with speed of $24 \, \text{km h}^{-1}$. The distance between the road sign and the post box is $240 \, \text{m}$. Find, in m s^{-2}, the deceleration of the car.

7 A skier is travelling downhill along a straight path with constant acceleration. At time $t = 0 \, \text{s}$, she passes a point A with speed $6 \, \text{m s}^{-1}$. She continues with the same acceleration until she reaches a point B with speed $15 \, \text{m s}^{-1}$. At B, the path flattens out and she travels from B to a point C at the constant speed of $15 \, \text{m s}^{-1}$. It takes $20 \, \text{s}$ for the skier to travel from B to C and the distance from A to C is $615 \, \text{m}$.

    a Sketch a speed–time graph to illustrate the motion of the skier.

    b Find the distance from A to B.

    c Find the time the skier took to travel from A to B.

8 A child drops a ball from a point at the top of a cliff which is $82 \, \text{m}$ above the sea. The ball is initially at rest. Find

    a the time taken for the ball to reach the sea,

    b the speed with which the ball hits the sea.

    c State one physical factor which has been ignored in making your calculation.

9 A particle moves along a straight line, from a point X to a point Y, with constant acceleration. The distance from X to Y is $104 \, \text{m}$. The particle takes $8 \, \text{s}$ to move from X to Y and the speed of the particle at Y is $18 \, \text{m s}^{-1}$. Find

    a the speed of the particle at X,

    b the acceleration of the particle.

    The particle continues to move with the same acceleration until it reaches a point Z. At Z the speed of the particle is three times the speed of the particle at X.

    c Find the distance XZ.

10 A pebble is projected vertically upwards with speed $21 \, \text{m s}^{-1}$ from a point $32 \, \text{m}$ above the ground. Find

    a the speed with which the pebble strikes the ground,

    b the total time for which the pebble is more than $40 \, \text{m}$ above the ground.

11 A particle P is moving along the x-axis with constant deceleration $2.5\ \text{m s}^{-2}$. At time $t = 0\ \text{s}$, P passes through the origin with velocity $20\ \text{m s}^{-1}$ in the direction of x increasing. At time $t = 12\ \text{s}$, P is at the point A. Find

    a the distance OA,   **b** the total distance P travels in $12\ \text{s}$.

12 A train starts from rest at a station P and moves with constant acceleration for $45\ \text{s}$ reaching a speed of $25\ \text{m s}^{-1}$. The train then maintains this speed for 4 minutes. The train then uniformly decelerates, coming to rest at a station Q.

    a Sketch a speed–time graph illustrating the motion of the train from P to Q.

    The distance between the stations is $7\ \text{km}$.

    b Find the deceleration of the train.

    c Sketch an acceleration–time graph illustrating the motion of the train from P to Q.

13

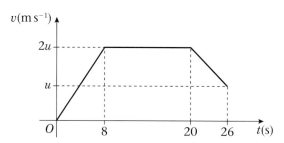

A particle moves $451\ \text{m}$ in a straight line. The diagram shows a speed–time graph illustrating the motion of the particle. The particle starts at rest and accelerates at a constant rate for $8\ \text{s}$ reaching a speed of $2u\ \text{m s}^{-1}$ at time $t = 26\ \text{s}$. Find

a the value of u,

b the distance moved by the particle while its speed is less than $u\ \text{m s}^{-1}$.

14 A particle is moving in a straight line. The particle starts with speed $5\ \text{m s}^{-1}$ and accelerates at a constant rate of $2\ \text{m s}^{-1}$ for $8\ \text{s}$. It then decelerates at a constant rate coming to rest in a further $12\ \text{s}$.

    a Sketch a speed–time graph illustrating the motion of the particle.

    b Find the total distance moved by the particle during its $20\ \text{s}$ of motion.

    c Sketch a distance–time graph illustrating the motion of the particle.

15 A boy projects a ball vertically upwards with speed $10\ \text{m s}^{-1}$ from a point X, which is $50\ \text{m}$ above the ground. T seconds after the first ball is projected upwards, the boy drops a second ball from X. Initially the second ball is at rest. The balls collide $25\ \text{m}$ above the ground. Find the value of T.

16 A car is moving along a straight road with uniform acceleration. The car passes a check-point A with speed $12\ \text{m s}^{-1}$ and another check-point C with speed $32\ \text{m s}^{-1}$. The distance between A and C is $1100\ \text{m}$.

    a Find the time taken by the car to move from A to C.

    Given that B is the mid-point of AC,

    b find the speed with which the car passes B.   **E**

17 A particle is projected vertically upwards with a speed of $30\,\text{m s}^{-1}$ from a point A. The point B is h metres above A. The particle moves freely under gravity and is above B for a time $2.4\,\text{s}$. Calculate the value of h.

18 Two cars A and B are moving in the same direction along a straight horizontal road. At time $t = 0$, they are side by side, passing a point O on the road. Car A travels at a constant speed of $30\,\text{m s}^{-1}$. Car B passes O with a speed of $20\,\text{m s}^{-1}$, and has constant acceleration of $4\,\text{m s}^{-2}$. Find

  a the speed of B when it has travelled $78\,\text{m}$ from O,

  b the distance from O of A when B is $78\,\text{m}$ from O,

  c the time when B overtakes A.

19 A car is being driven on a straight stretch of motorway at a constant speed of $34\,\text{m s}^{-1}$, when it passes a speed restriction sign S warning of road works ahead and requiring speeds to be reduced to $22\,\text{m s}^{-1}$. The driver continues at her speed for $2\,\text{s}$ after passing S. She then reduces her speed to $22\,\text{m s}^{-1}$ with constant deceleration of $3\,\text{m s}^{-2}$, and continues at the lower speed.

  a Draw a speed–time graph to illustrate the motion of the car after it passes S.

  b Find the shortest distance before the road works that S should be placed on the road to ensure that a car driven in this way has had its speed reduced to $22\,\text{m s}^{-1}$ by the time it reaches the start of the road works.

20 A train starts from rest at station A and accelerates uniformly at $3x\,\text{m s}^{-2}$ until it reaches a speed of $30\,\text{m s}^{-1}$. For the next T seconds the train maintains this constant speed. The train then retards uniformly at $x\,\text{m s}^{-2}$ until it comes to rest at a station B. The distance between the stations is $6\,\text{km}$ and the time taken from A to B is 5 minutes.

  a Sketch a speed–time graph to illustrate this journey.

  b Show that $\dfrac{40}{x} + T = 300$.

  c Find the value of T and the value of x.

  d Calculate the distance the train travels at constant speed.

  e Calculate the time taken from leaving A until reaching the point half-way between the stations.

Summary of key points.

1 You need to know these symbols and what they represent.

s	displacement (distance)
u	starting (initial) velocity
v	final velocity
a	acceleration
t	time

2 If a particle is slowing down it has a negative acceleration. This is called deceleration or retardation.

3 Convert all your measurements into base SI units before substituting values into the formulae.

Measurement	SI unit
time (t)	seconds (s)
displacement (s)	metres (m)
velocity (v or u)	metres per second (m s^{-1})
acceleration (a)	metres per second per second (m s^{-2})

4 You need to remember the five formulae for solving problems about particles moving in a straight line with constant acceleration.

- $v = u + at$
- $s = \left(\dfrac{u + v}{2}\right)t$
- $v^2 = u^2 + 2as$
- $s = ut + \frac{1}{2}at^2$
- $s = vt - \frac{1}{2}at^2$

5 An object moving vertically in a straight line can be modelled as a particle with a constant downward acceleration of $g = 9.8$ m s^{-2}.

6 The gradient of a speed–time graph illustrating the motion of a particle represents the acceleration of the particle.

7 The area under a speed–time graph illustrating the motion of a particle represents the distance moved by the particle.

8 Area of a trapezium = average of the parallel sides × height
$$= \tfrac{1}{2}(a + b) \times h$$

9 At constant speed, distance = speed × time

After completing this chapter you should be able to

- solve problems involving the forces acting on a body moving in a straight line with constant velocity or constant acceleration
- resolve a force into its components
- understand the coefficient of friction and solve problems involving friction
- solve problems involving collisions between bodies
- understand the Impulse – Momentum principle.

Dynamics of a particle moving in a straight line

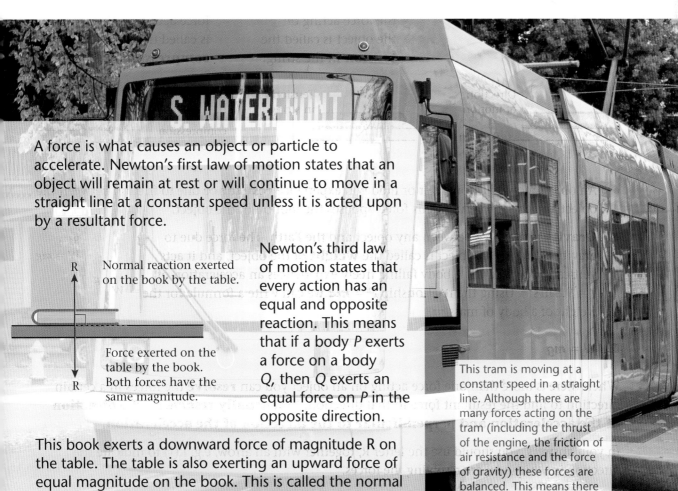

A force is what causes an object or particle to accelerate. Newton's first law of motion states that an object will remain at rest or will continue to move in a straight line at a constant speed unless it is acted upon by a resultant force.

R — Normal reaction exerted on the book by the table.

Force exerted on the table by the book. Both forces have the same magnitude. R

Newton's third law of motion states that every action has an equal and opposite reaction. This means that if a body P exerts a force on a body Q, then Q exerts an equal force on P in the opposite direction.

This book exerts a downward force of magnitude R on the table. The table is also exerting an upward force of equal magnitude on the book. This is called the normal reaction between the book and the table.

This tram is moving at a constant speed in a straight line. Although there are many forces acting on the tram (including the thrust of the engine, the friction of air resistance and the force of gravity) these forces are balanced. This means there is no resultant force

3.1 You can use Newton's Laws and the formula *F = ma* to solve problems involving force and acceleration.

■ Newton's second law of motion states that the force needed to accelerate a particle is equal to the product of the mass of the particle and the acceleration produced.

The unit of force is the **newton** (**N**). It is defined as the force that will cause a mass of 1 kg to accelerate at a rate of $1 \, \text{m s}^{-2}$.

■ ***F = ma***

This **equation of motion** gives a relationship between the resultant force, ***F*** newtons, acting on a particle of mass ***m*** kg producing an acceleration of ***a*** m s^{-2}.

You need to be able to understand the different types of force that can act on an object.

● The **normal reaction** is the force which acts perpendicular to a surface when an object is in contact with the surface. In the example on the previous page, the normal reaction is equal in magnitude to the weight of the book.

● **Friction** is a force which opposes the motion between two rough surfaces.	● If an object is being pulled along by a string, the force acting on the object is called the **tension** in the string.	● If an object is being pushed along using a light rod, the force acting on the object is called the **thrust** or **compression** in the rod.

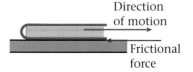

Direction of motion

Frictional force

Tension in string

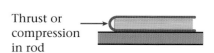

Thrust or compression in rod

● An object moving through air or fluid will experience **resistance** due to friction between the object and the air or fluid. This force opposes the motion of the object.

● **Gravity** is the force between any object and the Earth. The force due to gravity acting on an object is called the **weight** of the object, and it acts vertically downwards. A body falling freely experiences an acceleration of $g = 9.8 \, \text{m s}^{-2}$. Using the relationship ***F = ma*** we can write a formula for the weight of a body of mass ***m***.

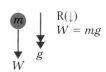

$R(\downarrow)$
$W = mg$

■ ***W = mg***

When there is more than one force acting on an object you can **resolve** the forces in a certain direction to find the resultant force in that direction. **You usually resolve in the direction of the acceleration and perpendicular to the direction of the acceleration.**

In your answers, you should use the letter R, together with an arrow, e.g. R(↑) to indicate the direction in which you are resolving the forces.

Example 1

Find the weight in newtons of a particle of mass 12 kg.

$W = mg$

$\quad = 12 \times 9.8$

$\quad = 117.6$

The particle weighs 120 N, correct to two significant figures.

Your value of g is correct to two significant figures. Give your answer to the same degree of accuracy.

Example 2

Find the acceleration when a particle of mass 1.5 kg is acted on by a resultant force of 6 N.

$F = ma$

$6 = 1.5a$

$a = 4$

The acceleration is $4 \, \text{m s}^{-2}$.

Substitute the values you know and solve the equation to find a.

Example 3

In each of these diagrams the body is accelerating as shown. Find the magnitudes of the unknown forces X and Y.

a

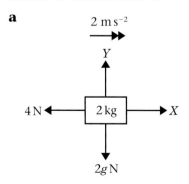

b

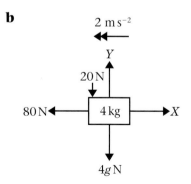

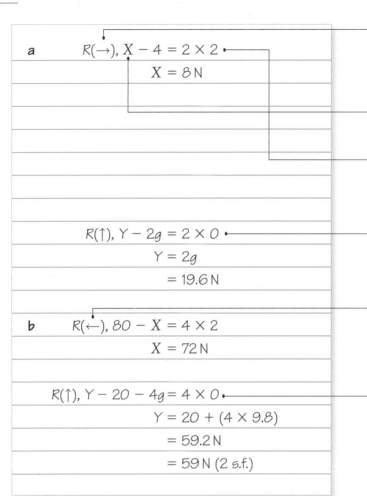

a $R(\rightarrow)$, $X - 4 = 2 \times 2$

   $X = 8\,\text{N}$

$R(\rightarrow)$, means apply $F = ma$ in the direction of the arrow i.e. the arrow is the positive direction.

On one side of the equation we include all the forces, giving each one the correct sign.

On the other side, we have mass $\times$ acceleration, giving the acceleration the correct sign.

   $R(\uparrow)$, $Y - 2g = 2 \times 0$

   $Y = 2g$

   $= 19.6\,\text{N}$

There is no vertical acceleration, so $a = 0$.

b $R(\leftarrow)$, $80 - X = 4 \times 2$

   $X = 72\,\text{N}$

It is usually easier to take the positive direction as the direction of the acceleration.

   $R(\uparrow)$, $Y - 20 - 4g = 4 \times 0$

   $Y = 20 + (4 \times 9.8)$

   $= 59.2\,\text{N}$

   $= 59\,\text{N}$ (2 s.f.)

There is no vertical acceleration, so $a = 0$.

Exercise 3A

Remember that g should be taken as $9.8\,\text{m s}^{-2}$.

1   Find the weight in newtons of a particle of mass $4\,\text{kg}$.

2   Find the mass of a particle whose weight is $490\,\text{N}$.

3   The weight of an astronaut on the Earth is $686\,\text{N}$. The acceleration due to gravity on the Moon is approximately $1.6\,\text{m s}^{-2}$. Find the weight of the astronaut when he is on the Moon.

4   Find the force required to accelerate a $1.2\,\text{kg}$ mass at a rate of $3.5\,\text{m s}^{-2}$.

5   Find the acceleration when a particle of mass $400\,\text{kg}$ is acted on by a resultant force of $120\,\text{N}$.

6   An object moving on a rough surface experiences a constant frictional force of $30\,\text{N}$ which decelerates it at a rate of $1.2\,\text{m s}^{-2}$. Find the mass of the object.

7 In each of the following scenarios, the forces acting on the body cause it to accelerate as shown. Find the magnitude of the unknown forces P and Q.

a

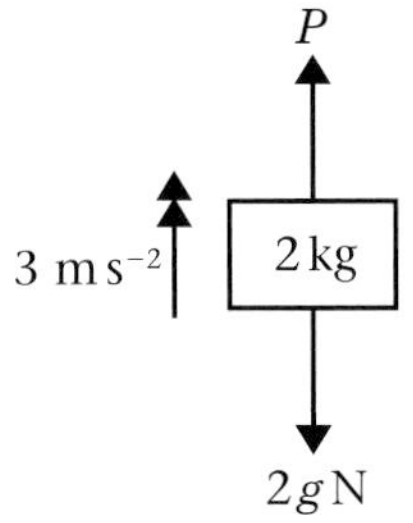

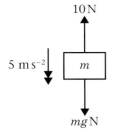

b

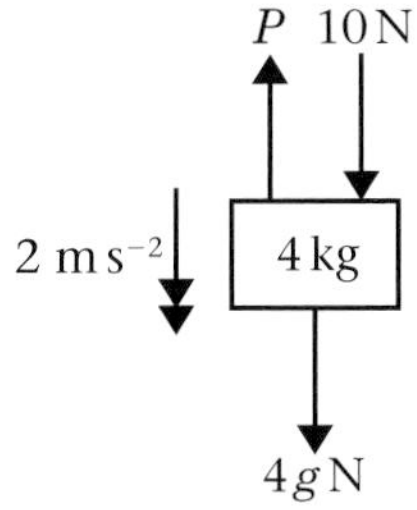

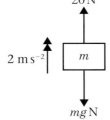

c

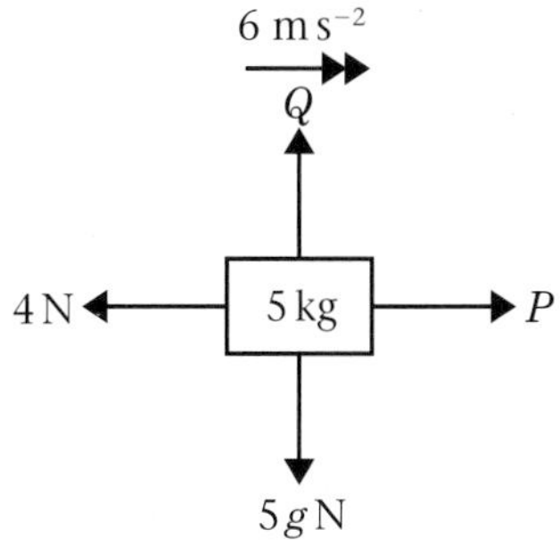

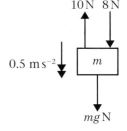

8 In each of the following situations, the forces acting on the body cause it to accelerate as shown. In each case find the mass of the body, m.

a

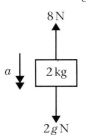

b

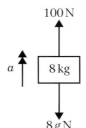

c

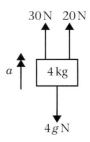

9 In each of the following situations, the forces acting on the body cause it to accelerate as shown with magnitude $a\,\mathrm{m\,s^{-2}}$. In each case find the value of a.

a

b

c

10 The diagram shows a block of mass 4 kg attached to a vertical rope.

Find the tension in the rope when the block moves downwards **a** with an acceleration of $2\,\mathrm{m\,s^{-2}}$, **b** at a constant speed of $4\,\mathrm{m\,s^{-1}}$, **c** with a deceleration of $0.5\,\mathrm{m\,s^{-2}}$.

3.2 You can solve problems involving forces by drawing a diagram showing all the relevant forces and resolving in one or more directions as necessary.

Example 4

A particle of mass 5 kg is pulled along a rough horizontal table by a horizontal force of magnitude 20 N against a constant friction force of magnitude 4 N. Given that the particle is initially at rest find

a the acceleration of the particle,

b the distance travelled by the particle in the first 4 seconds,

c the magnitude of the normal reaction between the particle and the table.

a

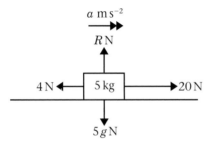

Draw a diagram showing all the forces and the acceleration.

$$R(\rightarrow),\ 20 - 4 = 5a$$

$$a = \frac{16}{5} = 3.2$$

The particle accelerates at $3.2\,\text{m s}^{-2}$.

Resolving horizontally, taking our positive direction as the direction of the acceleration, and using $F = ma$.

b $s = ut + \frac{1}{2}at^2$

$$s = (0 \times 4) + \frac{1}{2} \times 3.2 \times 4^2$$

$$= 25.6$$

The particle moves a distance of 25.6 m.

Since the acceleration is constant.

Substituting in the values.

c $R(\uparrow),\ R - 5g = 5 \times 0 = 0$

$$R = 5g = 5 \times 9.8 = 49\,\text{N}$$

The normal reaction has magnitude 49 N.

Resolving vertically, taking up as positive. Write all the forces on the left-hand-side of the equation and mass × acceleration on the right-hand-side.

Example 5

A small pebble of mass 500 g is attached to the lower end of a light string. Find the tension in the string when the pebble

a is moving upwards with an acceleration of $2\,\mathrm{m\,s^{-2}}$,

b is moving downwards with an acceleration of $3\,\mathrm{m\,s^{-2}}$,

c is moving downwards at a constant speed of $5\,\mathrm{m\,s^{-1}}$.

d is moving downwards with a deceleration of $4\,\mathrm{m\,s^{-2}}$.

a

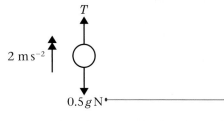

$R(\uparrow),\ T - 0.5g = 0.5 \times 2$

$T = 5.9\,\mathrm{N}$

Draw a diagram showing all the forces and the acceleration.	

Convert 500 g into 0.5 kg first.

Resolving vertically upwards, in the direction of the acceleration.

b

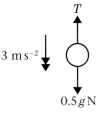

$R(\downarrow),\ 0.5g - T = 0.5 \times 3$

$T = 3.4\,\mathrm{N}$

Draw a diagram showing the forces and acceleration.

Resolving vertically downwards, in the direction of the acceleration.

c

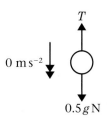

$R(\downarrow),\ 0.5g - T = 0.5 \times 0$

$T = 4.9\,\mathrm{N}$

Constant speed means zero acceleration.

Deceleration is shown as negative acceleration.

d

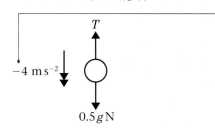

$R(\downarrow),\ 0.5g - T = 0.5 \times -4$

$T = 6.9\,\mathrm{N}$

Resolving, taking downwards as positive.

Exercise 3B

1 A ball of mass 200 g is attached to the upper end of a vertical light rod. Find the thrust in the rod when it raises the ball vertically with an acceleration of $1.5\,\text{m s}^{-2}$.

2 A small pebble of mass 50 g is dropped into a pond and falls vertically through it with an acceleration of $2.8\,\text{m s}^{-2}$. Assuming that the water produces a constant resistance, find its magnitude.

3 A lift of mass 500 kg is lowered or raised by means of a metal cable attached to its top. The lift contains passengers whose total mass is 300 kg. The lift starts from rest and accelerates at a constant rate, reaching a speed of $3\,\text{m s}^{-1}$ after moving a distance of 5 m. Find

   a the acceleration of the lift,

   b the tension in the cable if the lift is moving vertically downwards,

   c the tension in the cable if the lift is moving vertically upwards.

4 A block of mass 1.5 kg falls vertically from rest and hits the ground 16.6 m below after falling for 2 s. Assuming that the air resistance experienced by the block as it falls is constant, find its magnitude.

5 A trolley of mass 50 kg is pulled from rest in a straight line along a horizontal path by means of a horizontal rope attached to its front end. The trolley accelerates at a constant rate and after 2 s its speed is $1\,\text{m s}^{-1}$. As it moves, the trolley experiences a resistance to motion of magnitude 20 N. Find

   a the acceleration of the trolley,   b the tension in the rope.

6 A trailer of mass 200 kg is attached to a car by a light tow-bar. The trailer is moving along a straight horizontal road and decelerates at a constant rate from a speed of $15\,\text{m s}^{-1}$ to a speed of $5\,\text{m s}^{-1}$ in a distance of 25 m. Assuming there is no resistance to the motion, find

   a the deceleration of the trailer,   b the thrust in the tow-bar.

7 A woman of mass 60 kg is in a lift which is accelerating upwards at a rate of $2\,\text{m s}^{-2}$.

   a Find the magnitude of the normal reaction of the floor of the lift on the woman.

   The lift then moves at a constant speed and then finally decelerates to rest at $1.5\,\text{m s}^{-2}$.

   b Find the magnitude of the normal reaction of the floor of the lift on the woman during the period of deceleration.

   c Hence explain why the woman will feel heavier during the period of acceleration and lighter during the period of deceleration.

8 The engine of a van of mass 400 kg cuts out when it is moving along a straight horizontal road with speed $16\,\text{m s}^{-1}$. The van comes to rest without the brakes being applied.

   In a model of the situation it is assumed that the van is subject to a resistive force which has constant magnitude of 200 N.

   a Find how long it takes the van to stop.

   b Find how far the van travels before it stops.

   c Comment on the suitability of the modelling assumption.

9 Albert and Bella are both standing in a lift. The mass of the lift is 250 kg. As the lift moves upward with constant acceleration, the floor of the lift exerts forces of magnitude 678 N and 452 N respectively on Albert and Bella. The tension in the cable which is pulling the lift upwards is 3955 N.

  a Find the acceleration of the lift.

  b Find the mass of Albert.

  c Find the mass of Bella.

10 A small stone of mass 400 g is projected vertically upwards from the bottom of a pond full of water with speed 10 m s^{-1}. As the stone moves through the water it experiences a constant resistance of magnitude 3 N. Assuming that the stone does not reach the surface of the pond, find

  a the greatest height above the bottom of the pond that the stone reaches,

  b the speed of the stone as it hits the bottom of the pond on its return,

  c the total time taken for the stone to return to its initial position on the bottom of the pond.

3.3 **If a force is applied at an angle to the direction of motion you can resolve it to find the component of the force that acts in the direction of motion.**

This book is being dragged along the table by means of a force of magnitude F. The book is moving horizontally, and the angle between the force and the direction of motion is θ.

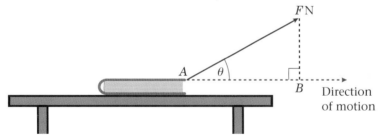

The effect of the force in the direction of motion is the length of the line AB. This is called the **component of the force in the direction of motion**. Using the rule for a right-angled triangle $\cos \theta = \dfrac{adjacent}{hypotenuse}$ you can see that the length of AB is $F \times \cos \theta$. Finding this value is called **resolving** the force in the direction of motion.

■ **The component of a force of magnitude F in a certain direction is $F \cos \theta$, where θ is the size of the angle between the force and the direction.**

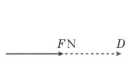

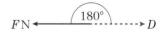

If F acts in the direction D, then the component of F in that direction is $F \cos 0° = F \times 1 = F$

If F acts at the right angles to D, then the component of F in that direction is $F \cos 90° = F \times 0 = 0$

If F acts in the opposite direction to D, then the component of F in that direction is $F \cos 180° = F \times -1 = -F$

Example 6

Find the component of each force in **i** the *x*-direction, **ii** the *y*-direction.

a

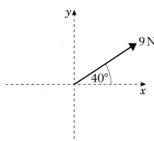

b

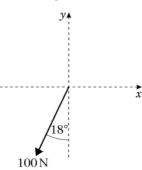

a **i** $\theta = 40°$

Component in *x*-direction = $F \cos \theta$

$= 9 \times \cos 40°$

$= 6.89 \, N \, (3 \text{ s.f.})$

> Give your answers correct to three significant figures.

ii

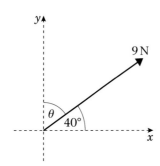

$\theta = 90° - 40°$

$= 50°$

> Make sure you find the angle between the force and the direction you are resolving in.

Component in *y*-direction = $F \cos \theta$

$= 9 \times \cos 50°$

$= 5.79 \, N \, (3 \text{ s.f.})$

> You are resolving in the positive *x*-direction so your answer is going to be negative.

b **i**

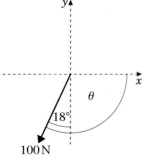

$\theta = 90° + 18°$

$= 108°$

Component in *x*-direction = $F \cos \theta$

$= 100 \times \cos 108°$

$= -30.9 \, N \, (3 \text{ s.f.})$

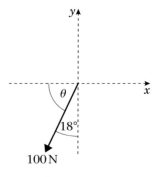

> You could resolve in the negative *x*-direction using $\theta = 90° - 18°$, then change the sign of your answer from positive to negative, i.e. component
> $= -100 \cos 72°$
> $= -30.9 \, N \, (3 \text{ s.f.})$

ii

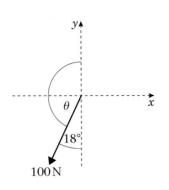

$\theta = 180° - 18°$

$= 162°$ •

> You could use $\theta = 18°$ then change the sign of your answer from positive to negative, i.e. component
> $= -100 \cos 18°$
> $= -95.1\,N$ (3 s.f.).

Component in y-direction $= F \cos \theta$

$= 100 \times \cos 162°$

$= -95.1\,N$ (3 s.f.)

Exercise 3C

1 Find the component of each force in
i the x-direction, **ii** the y-direction.

a

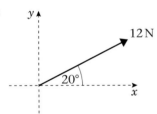

b

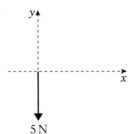

c

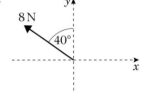

d

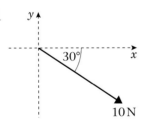

e

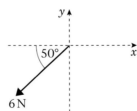

f

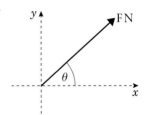

2 For each of the following systems of forces, find the sum of the components in
i the x-direction, **ii** the y-direction.

a

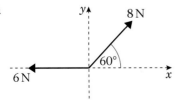

b

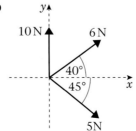

c

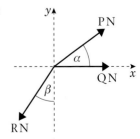

3.4 You can calculate the magnitude of a frictional force using the coefficient of friction.

Friction is a force which opposes motion between two rough surfaces. It occurs when the two surfaces are moving relative to one another, or when there is a **tendency** for them to move relative to one another.

This block is stationary. There is no horizontal force being applied, so there is no tendency for the block to move. There is no frictional force acting on the block.

This block is also stationary. There is a horizontal force being applied which is not sufficient to move the block. There is a tendency for the block to move, but it doesn't because the force of friction is equal and opposite to the force being applied.

As the applied force increases, the force of friction increases to prevent the block from moving. If the magnitude of the applied force exceeds a certain **maximum** or **limiting value**, the block will move. While the block moves, the force of friction will remain constant at its maximum value.

The limiting value of the friction depends on two things:
  ○ the normal reaction R between the two surfaces in contact,
  ○ the roughness of the two surfaces in contact.

You can measure roughness using the **coefficient of friction**, which is represented by the letter μ (pronounced *myoo*). The rougher the two surfaces are, the larger the value of μ. For smooth surfaces there is no friction and $\mu = 0$.

■ The maximum or limiting value of the friction F_{MAX} between two surfaces is given by

$$F_{MAX} = \mu R$$

where μ is the coefficient of friction and R is the normal reaction between the two surfaces.

Example 7

Find the maximum frictional force which can act on a block of mass 4 kg which rests on a rough horizontal plane, if the coefficient of friction between the block and the plane is **a** 0.2, **b** 0.7.

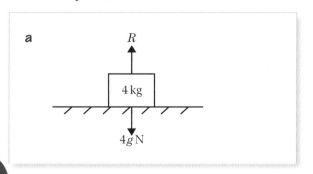

First draw a diagram showing all the forces acting on the block.

$$R(\uparrow), \quad R - 4g = 4 \times 0 = 0$$

There is no acceleration upwards.

$$R = 39.2 \, N$$

$$F_{MAX} = \mu R = 0.2 \times 39.2$$

You can now calculate the maximum 'available' friction force.

$$= 7.84 = 7.8 \, N \, (2 \, s.f.)$$

b As before, $R = 39.2 \, N$

You now have a different value for μ.

$$F_{MAX} = \mu R = 0.7 \times 39.2$$

$$= 27.44 = 27 \, N \, (2 \, s.f.)$$

Since μ is higher there is more friction 'available'.

Example 8

A block of mass 5 kg lies at rest on rough horizontal ground. The coefficient of friction between the block and the ground is 0.4. A horizontal force P is applied to the block. Find the magnitude of the friction force acting on the block and the acceleration of the block when the magnitude of P is
a 10 N, **b** 19.6 N, **c** 30 N.

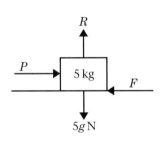

First draw a diagram showing all the forces acting on the block.

You then need to calculate what the maximum possible friction force is in this situation.

$$R(\uparrow), \quad R = 5g = 49 \, N$$

The normal reaction will equal the weight as there is no acceleration upwards.

So

$$F_{MAX} = \mu R = 0.4 \times 49$$

$$= 19.6 \, N$$

The maximum available friction force is 19.6 N.

Do not round this value as you will need to use it in your calculations.

a When $P = 10 \, N$, the friction will only need to be 10 N to prevent the block from sliding and the block will remain at rest in equilibrium.

b When $P = 19.6 \, N$, the friction will need to be at its maximum value of 19.6 N to prevent the block from sliding, and the block will remain at rest in **limiting equilibrium**.

c When $P = 30 \, N$, the friction will be unable to prevent the block from sliding, but it will remain at its maximum value of 19.6 N. The block will accelerate from rest along the plane in the direction of P with acceleration a.

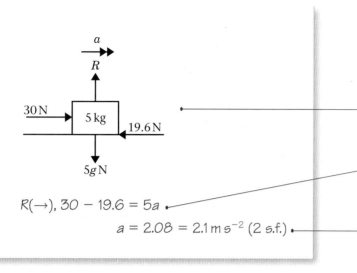

Draw a diagram showing all the forces and the acceleration.

Resolving in the direction of the acceleration.

$R(\rightarrow)$, $30 - 19.6 = 5a$

The answer is rounded to two significant figures as we have used $g = 9.8\,\mathrm{m\,s^{-2}}$.

$a = 2.08 = 2.1\,\mathrm{m\,s^{-2}}$ (2 s.f.)

■ If a force P is applied to a block of mass m which is at rest on a rough horizontal surface and P acts at an angle to the horizontal:

- the normal reaction R is not equal to mg,
- the force tending to pull or push the block along the plane is not equal to P.

Example 9

A 5 kg box lies at rest on a rough horizontal floor. The coefficient of friction between the box and the floor is 0.5. A force P is applied to the box to pull or push it horizontally along the floor. Find the magnitude of P which is necessary to achieve this if

a P is applied horizontally,

b P is applied at an angle α above the horizontal, where $\tan \alpha = \frac{3}{4}$,

c P is applied at an angle α below the horizontal, where $\tan \alpha = \frac{3}{4}$.

a

Draw a diagram showing the forces acting.

There is no acceleration upwards.

$R(\uparrow)$, $\quad R - 5g = 5 \times 0 = 0$

$R = 5g = 49\,\mathrm{N}$

∴ $\quad F_{\mathrm{MAX}} = \mu R = 0.5 \times 49$

Find the maximum possible friction force.

$= 24.5\,\mathrm{N}$

In the position of limiting equilibrium there is no acceleration, so $P = F_{\mathrm{MAX}} = 24.5\,\mathrm{N}$

P must exceed 24.5 N.

b Note that if $\tan \alpha = \dfrac{3}{4}$,

then $\sin \alpha = \dfrac{3}{5} = 0.6$

and $\cos \alpha = \dfrac{4}{5} = 0.8$.

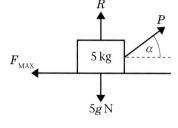

Draw a new diagram.

Again there is no acceleration upwards.

$\cos (90° - \alpha) = \sin \alpha$
$= 0.6$

$R(\uparrow), \quad R + P \cos (90° - \alpha) - 5g = 0$

$R = 49 - P \sin \alpha = 49 - 0.6P$

No acceleration horizontally.

$R(\rightarrow), \quad P \cos \alpha - F_{MAX} = 0$

$P \cos \alpha = F_{MAX}$

Use the expression for R found above.

$F_{MAX} = \mu R = 0.5(49 - 0.6P)$

Use $\cos \alpha = 0.8$.

So $0.8P = 0.5(49 - 0.6P)$

$0.8P = 24.5 - 0.3P$

Multiply out.

$1.1P = 24.5$

Collect the P terms.

$P = 22 \text{ N (2 s.f.)}$

Solve for P.

P must exceed 22 N.

Note that this is less than the value found in part **a**.

c

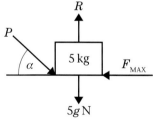

Draw another diagram.

$R(\uparrow), R - 5g - P \cos (90° - \alpha) = 5 \times 0$

$= 0$

There is no acceleration vertically.

$R = 5g + P \sin \alpha$

$= 49 + 0.6P$

Use $\cos (90° - \alpha) = \sin \alpha$ as before.

$F_{MAX} = \mu R = 0.5(49 + 0.6P)$

$= 24.5 + 0.3P$

Use the expression for R found above.

$R(\rightarrow), P \cos \alpha - F_{MAX} = 5 \times 0 = 0$

No acceleration horizontally.

$0.8P = 24.5 + 0.3P$

Substituting for F_{MAX} from above.

$0.5P = 24.5$

$P = 49 \text{ N}$

P must exceed 49 N.

Note the size of this answer compared with **b**.

Exercise 3D

1 Each of the following diagrams shows a body of mass 5 kg lying initially at rest on rough horizontal ground. The coefficient of friction between the body and the ground is $\frac{1}{7}$. In each diagram R is the normal reaction from the ground on the body and F is the friction force exerted on the body by the ground. Any other forces applied to the body are as shown on the diagram. In each case

i find the magnitude of F,

ii state whether the body will remain at rest or accelerate from rest along the ground,

iii find, when appropriate, the magnitude of this acceleration.

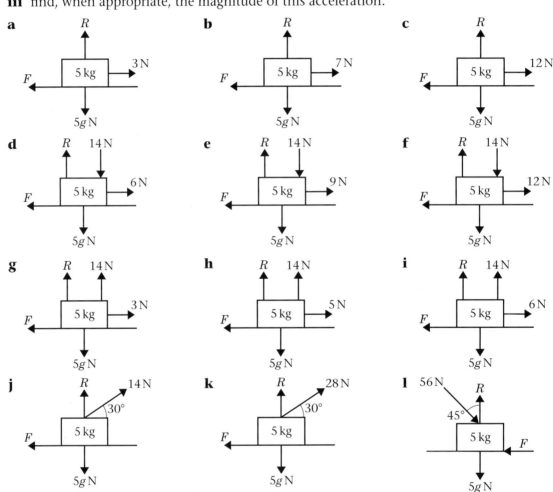

2 In each of the following diagrams, the forces shown cause the body of mass 10 kg to accelerate as shown along the rough horizontal plane. R is the normal reaction and F is the friction force. Find the coefficient of friction in each case.

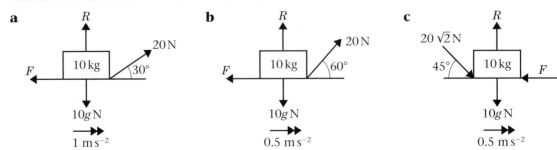

3.5 You can solve problems about a particle on an inclined plane by resolving the forces parallel and perpendicular to the plane.

■ If a particle of mass m is placed on a smooth inclined plane and released it will slide down the slope.

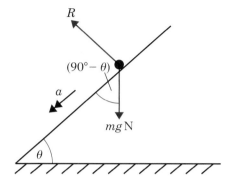

Resolve in the direction of motion, which is down the slope.

$$R(\nearrow) \quad \cancel{m}g \cos (90° - \theta) = \cancel{m}a$$
$$g \cos (90° - \theta) = a$$
$$g \sin \theta = a$$

The mass of the particle does not affect the acceleration, but the angle of the slope does.

If the plane is rough, the force of friction might be sufficient to prevent the particle from moving.

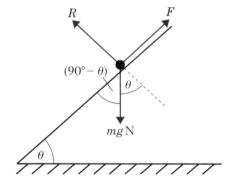

Resolve perpendicular to the plane. Remember that the normal reaction acts at right-angles to the plane.

$$R(\nwarrow) \quad R - mg \cos \theta = m \times 0 = 0$$
$$R = mg \cos \theta$$

Now resolve in the direction of the slope. If the particle is stationary:

$$R(\nearrow) \quad mg \cos (90° - \theta) - F = m \times 0 = 0$$
$$mg \sin \theta = F$$

The frictional force F is always less than or equal to F_{MAX}.

$$F \leqslant \mu R$$
$$\cancel{m}g \sin \theta \leqslant \mu \cancel{m}g \cos \theta$$
$$\frac{\sin \theta}{\cos \theta} \leqslant \mu$$
$$\tan \theta \leqslant \mu$$

■ A particle placed on a rough inclined plane will remain at rest if $\tan \theta \leqslant \mu$, where θ is the angle the plane makes with the horizontal and μ is the coefficient of friction between the particle and the plane.

If $\tan \theta > \mu$ then the particle will accelerate down the slope.

Example 10

A particle is held at rest on a rough plane which is inclined to the horizontal at an angle α, where $\tan \alpha = 0.75$. The coefficient of friction between the particle and the plane is 0.5. The particle is released and slides down the plane. Find

a the acceleration of the particle,

b the distance it slides in the first 2 seconds.

a

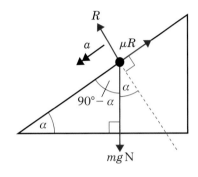

Draw a diagram showing all the forces and the acceleration. Note that you are not given the mass of the particle so call it m.

Since the particle slides down the plane friction will be limiting, so $\mu R = 0.5R$

$R(\nwarrow)$, $\qquad R - mg \cos \alpha = m \times 0 = 0$

Resolving perpendicular to the acceleration.

$\qquad\qquad\qquad R = mg \cos \alpha \qquad$ ①

$R(\nearrow)$, $\qquad mg \sin \alpha - \mu R = ma \qquad$ ②

Resolving in the direction of the acceleration.

From equation ①,

$R = 0.8mg$

If $\tan \alpha = 0.75 = \frac{3}{4}$,

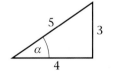

Then equation ② becomes

$0.6\cancel{m}g - 0.5 \times (0.8\cancel{m}g) = \cancel{m}a$

$\qquad 0.6g - 0.4g = a$

$\qquad\qquad 0.2g = a$

$\cos \alpha = \frac{4}{5} = 0.8$

and $\sin \alpha = \frac{3}{5} = 0.6$

(or use your calculator)

The acceleration of the particle is $0.2g$ or 1.96 m s^{-2} (3 s.f.) or 2.0 m s^{-2} (2 s.f.)

Substitute for $\sin \alpha$, μ and R and cancel the m's.

b Since the acceleration is *constant* we can use a formula:

Any of these answers would be acceptable.

$u = 0, a = 0.2g, t = 2, s = ?$

$s = ut + \frac{1}{2}at^2$

$s = 0 + \frac{1}{2} \times 0.2g \times 2^2$

Choose the appropriate formula.

$\quad = 3.92 = 3.9$ (2 s.f.)

The particle slides 3.9 m (2 s.f.) down the plane.

Substitute in the values.

Example 11

A box of mass 2 kg is pushed up a rough plane by a horizontal force of magnitude 25 N. The plane is inclined to the horizontal at an angle of 10°. Given that the coefficient of friction between the box and the plane is 0.3, find the acceleration of the box.

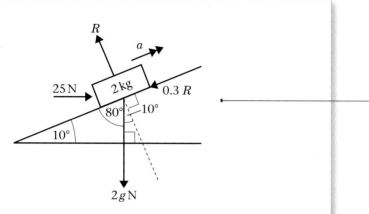

Draw a diagram showing all forces and the acceleration. Note that friction will be limiting, since the box is sliding.

$R(\nwarrow)$, $R - 2g \cos 10° - 25 \cos 80° = 2 \times 0 = 0$

$\qquad R = 2g \cos 10° + 25 \cos 80°$ ①

Resolving perpendicular to the acceleration.

$R(\nearrow)$, $25 \cos 10° - 2g \cos 80° - 0.3R = 2a$

Resolving in the direction of the acceleration.

$\quad 25 \cos 10° - 2g \cos 80°$
$\quad - 0.3 (2g \cos 10° + 25 \cos 80°) = 2a$

$\quad (25 - 0.6g) \cos 10°$
$\quad - (2g + 7.5) \cos 80° = 2a$

$\qquad 14.124 \ldots \qquad = 2a$

$\qquad a \quad = 7.1 \, \mathrm{m \, s^{-2}} \, (2 \, \text{s.f.})$

Substituting for R from equation ① and simplifying.

The box accelerates up the plane at $7.1 \, \mathrm{m \, s^{-2}} \, (2 \, \text{s.f.})$.

Exercise 3E

1 A particle of mass 0.5 kg is placed on a smooth inclined plane. Given that the plane makes an angle of 20° with the horizontal, find the acceleration of the particle.

2 The diagram shows a box of mass 2 kg being pushed up a smooth plane by a horizontal force of magnitude 20 N. The plane is inclined to the horizontal at an angle α, where $\tan \alpha = \frac{3}{4}$.

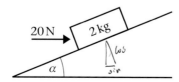

Find

a the normal reaction between the box and the plane,

b the acceleration of the box up the plane.

3 A boy of mass 40 kg slides from rest down a straight slide of length 5 m. The slide is inclined to the horizontal at an angle of 20°. The coefficient of friction between the boy and the slide is 0.1. By modelling the boy as a particle, find

a the acceleration of the boy,

b the speed of the boy at the bottom of the slide.

4 A block of mass 20 kg is released from rest at the top of a rough slope. The slope is inclined to the horizontal at an angle of 30°. After 6 s the speed of the block is 21 m s⁻¹. As the block slides down the slope it is subject to a constant resistance of magnitude R N. Find the value of R.

5 A book of mass 2 kg slides down a rough plane inclined at 20° to the horizontal. The acceleration of the book is 1.5 m s⁻². Find the coefficient of friction between the book and the plane.

6 A block of mass 4 kg is pulled up a rough slope, inclined at 25° to the horizontal, by means of a rope. The rope lies along the line of the slope. The tension in the rope is 30 N. Given that the acceleration of the block is 2 m s⁻² find the coefficient of friction between the block and the plane.

7 A parcel of mass 10 kg is released from rest on a rough plane which is inclined at 25° to the horizontal.

a Find the normal reaction between the parcel and the plane.

Two seconds after being released the parcel has moved 4 m down the plane.

b Find the coefficient of friction between the parcel and the plane.

8 A particle P is projected up a rough plane which is inclined at an angle α to the horizontal, where $\tan \alpha = \frac{3}{4}$. The coefficient of friction between the particle and the plane is $\frac{1}{3}$. The particle is projected from the point A with speed 20 m s⁻¹ and comes to instantaneous rest at the point B.

a Show that while P is moving up the plane its deceleration is $\dfrac{13g}{15}$.

b Find, to three significant figures, the distance AB.

c Find, to three significant figures, the time taken for P to move from A to B.

d Find the speed of P when it returns to A.

3.6 **You can solve problems involving connected particles by considering the particles separately.**

If a system involves the motion of more than one particle, the particles may be considered separately. Particular care is then needed to ensure that all the forces acting on each particle are considered.

Provided that all parts of the system are moving in the **same straight line**, then you can also treat the whole system as a single particle.

Example 12

Two particles P and Q, of masses 5 kg and 3 kg respectively, are connected by a light inextensible strong. Particle P is pulled by a horizontal force of magnitude 40 N along a rough horizontal plane. The coefficient of friction between each particle and the plane is 0.2. The string is taut.

a Find the acceleration of each particle.

b Find the tension in the string.

c Explain how the modelling assumptions that the string is light and inextensible have been used.

a

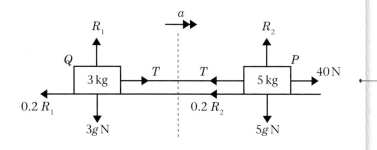

Each mass will have a different normal reaction and therefore a different friction.

For Q: R($\uparrow$), $R_1 - 3g = 0$

$\qquad\qquad R_1 = 3g$ N

No acceleration upwards.

For P: R($\uparrow$), $R_2 - 5g = 0$

$\qquad\qquad R_2 = 5g$ N

For the whole system (P and Q):

R($\rightarrow$), $40 - (0.2 \times 3g) - (0.2 \times 5g) = 8a$

$\qquad\qquad 40 - 1.6g = 8a$

$\qquad\qquad\qquad a = 3.04 \, \text{m s}^{-2}$

You can do this because both masses are moving in the same straight line.

The tension in the string is not included as you are treating the whole system as a single particle.

The acceleration of each mass is $3.04 \, \text{m s}^{-2}$ since the string is inextensible.

b For Q: R($\rightarrow$), $T - 0.2 \times 3g = 3 \times 3.04$

$\qquad\qquad\qquad T = 15$ N

To find T consider one of the particles. It is easier to consider Q rather than P.

The tension in the string is 15 N.

c Inextensible $\Rightarrow$ acceleration of masses is the same.

$\qquad$ light $\Rightarrow$ tension is the same throughout the length of the string and the mass of the string is negligible.

Example **13**

A light scale-pan is attached to a vertical light inextensible string. The scale-pan carries two masses A and B. The mass of A is 400 g and the mass of B is 600 g. A rests on top of B, as shown in the diagram.

The scale-pan is raised vertically, using the string, with acceleration 0.5 m s^{-2}.

a Find the tension in the string.

b Find the force exerted on mass B by mass A.

c Find the force exerted on mass B by the scale-pan.

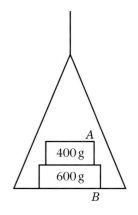

a

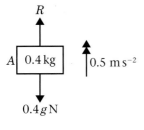

For the whole system:

R(↑) $T - 0.4g - 0.6g = (0.4 + 0.6)a$

so, $T - g = 1 \times 0.5$

                          $T = 10.3 \text{ N}$

The tension in the string is 10.3 N (3 s.f.) or 10 N (2 s.f.).

You can use this since all parts of the system are moving in the same straight line.

Note that we must convert 400 g to 0.4 kg and 600 g to 0.6 kg.

$a = 0.5$.

Simplify.

b

R

A | 0.4 kg

0.5 m s^{-2}

$0.4g \text{ N}$

For A only:

R(↑) $R - 0.4g = 0.4 \times 0.5$

                 $R = 4.12 \text{ N}$

So the force exerted on B by A is 4.12 N (3 s.f.) or 4.1 N (2 s.f.).

We find the force exerted on A by B and then use Newton's 3rd Law to say that the force exerted on B by A will have the same magnitude.

c Method 1

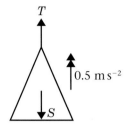

For scale-pan only.

$R(\uparrow)$, $T - S = 0 \times 0.5$

$\qquad = 0$

So $\qquad T = S = 10.3\,\text{N}$

So, the force exerted on B by the scale-pan is 10.3 N upwards

> It's easier to find the force exerted on the scale-pan by B and then use Newton's 3rd Law to say that the force exerted on B by the scale-pan has the same magnitude but is in the opposite direction.

> The scale-pan is light i.e. its mass is 0.

> From part **a**.

> Use Newton's 3rd Law.

Method 2

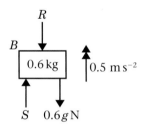

Consider B only.

$R(\uparrow)$, $S - R - 0.6g = 0.6 \times 0.5$

$\qquad S - 4.12 - 5.88 = 0.3$

$\qquad\qquad S = 10.3\,\text{N}$

> Substitute for R from **b**.

Method 3

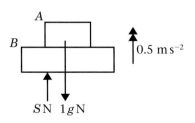

Consider A and B only.

$R(\uparrow)$, $\quad S - g = 1 \times 0.5$

$\qquad S - 9.8 = 0.5$

$\qquad\qquad S = 10.3\,\text{N}$

> Total mass of A and B is 1 kg.

■ In problems involving particles which are connected by string(s) which pass over pulley(s) you cannot treat the whole system as a single particle. This is because the particles are moving in different directions.

Example 14

Particles P and Q, of masses $2m$ and $3m$, are attached to the ends of a light inextensible string. The string passes over a small smooth fixed pulley and the masses hang with the string taut. The system is released from rest.

a Find the acceleration of each mass.

b Comment on any modelling assumptions used.

c Find the tension in the string.

d Find the force exerted on the pulley by the string.

e Find the distance moved by Q in the first 4 s, assuming that P does not reach the pulley.

a

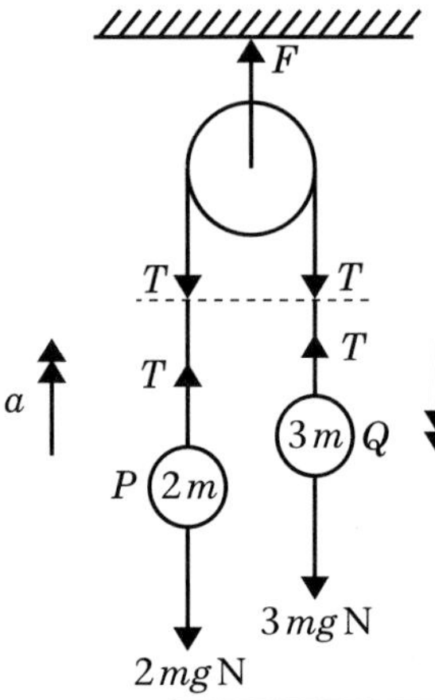

Draw a diagram showing all the forces acting on each mass and the pulley, and the acceleration.

For P: R(↑), $T - 2mg = 2ma$ ①
For Q: R(↓), $3mg - T = 3ma$ ②

Now resolve for each mass separately, in the direction of its acceleration.

Adding equations ① and ②:

$3mg - \cancel{T} + \cancel{T} - 2mg = 3ma + 2ma$

Add the equations to eliminate T.

$\cancel{m}g = 5\cancel{m}a$

Simplifying.

$\frac{1}{5}g = a$

The acceleration of each mass is $\frac{1}{5}g$

$(1.96 \, \text{m s}^{-2} \text{ or } 2.0 \, \text{m s}^{-2} \, (2 \text{ s.f.}))$

Any of these answers would be acceptable.

b The lightness of the string means that the tension in it is constant.

The inextensibility of the string means that both masses have the same acceleration.

The 'small smooth' pulley means that the tension will be the same on both sides of the pulley.

c From ①, $T - 2mg = 2m \times \frac{1}{5}g$ •————————— Substitute for a.

$$T = \frac{12mg}{5} \text{ N} •$$ ————————— Collect terms.

The tension in the string is $\frac{12mg}{5}$ N.

d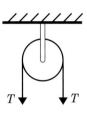

The force exerted on the pulley by the string is $2T$ N downwards or $\frac{24mg}{5}$ N.

$T \downarrow \qquad \downarrow T$

e $u = 0$, $a = \frac{1}{5}g$, $t = 4$, $s = ?$

$$s = ut + \frac{1}{2}at^2 •$$ ————————— Since a is a constant we can use any of the formulae for constant acceleration.

$$= 0 + \frac{1}{2} \times 1.96 \times 4^2$$

$$= 15.68 \text{ m}$$

$$= 15.7 \text{ m (3 s.f.)} •$$ ————————— Either of those answers would be acceptable.

Q moves through a distance of 16 m (2 s.f.). •

Example 15

Two particles A and B of masses 0.4 kg and 0.8 kg respectively are connected by a light inextensible string. Particle A lies on a rough horizontal table 4.5 m from a small smooth pulley which is fixed at the edge of the table. The string passes over the pulley and B hangs freely, with the string taut, 0.5 m above horizontal ground. The coefficient of friction between A and the table is 0.2. The system is released from rest. Find

a the acceleration of the system,

b the time taken for B to reach the ground,

c the total distance travelled by A before it first comes to rest.

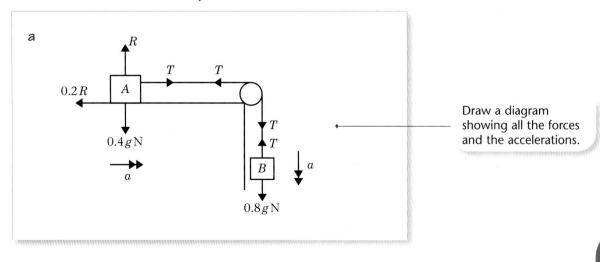

Draw a diagram showing all the forces and the accelerations.

For A only: R(↑), $R - 0.4g = 0$ •————————— No acceleration upwards.

$$R = 0.4g\,\text{N}$$

R(→), $T - 0.2R = 0.4a$ •————————— Resolve in the direction of a.

$$T - 0.08g = 0.4a \qquad ① •$$ ————— Substitute for R.

For B only: R(↓), $0.8g - T = 0.8a \qquad ②$

Add ① and ②, •————————————————— Resolve in the direction of the acceleration.

$$0.8g - \cancel{T} + \cancel{T} - 0.08g = 0.8a + 0.4a$$

$$0.72g = 1.2a$$ ————————————————— To eliminate the T terms.

$$0.6g = a$$

The acceleration of the system is $0.6g$. •———— Any of these answers is acceptable.

or $5.88\,\text{m s}^{-2}$ (3 s.f.) or $5.9\,\text{m s}^{-2}$ (2 s.f.)

b

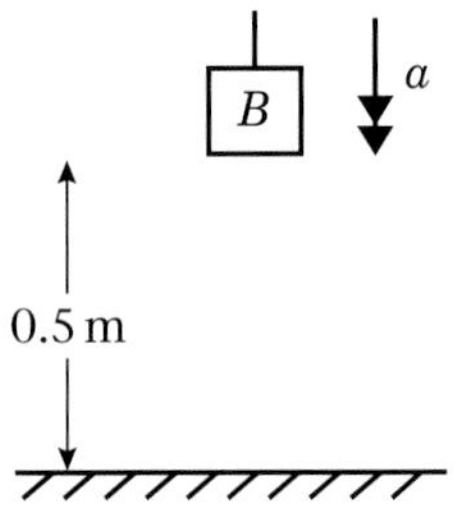

$u = 0, s = 0.5,$ •————————— Note we are using an unrounded value of the acceleration.

$a = 5.88,\ t = ?$

$s = ut + \frac{1}{2}at^2$ •————————— The acceleration is constant.

$0.5 = 0 + \frac{1}{2} \times 5.88 \times t^2$

$t = 0.412$ (3 s.f.)

The time taken for B to hit the ground is $0.41\,\text{s}$ (2 s.f.)

c We need to find the speed of B when it hits the ground.

$u = 0, a = 5.88, t = 0.41239, v = ?$

$v = u + at$

$v_B = 0 + 5.88 \times 0.41239$ •————————— Note we are using an unrounded value for t.

$= 2.42487\,\text{m s}^{-1}.$

So the speed of A on the table is $2.42487\,\text{m s}^{-1}$ •——— Using surds, $v_B = \sqrt{\dfrac{3g}{5}}$

Once B hits the ground the string will go slack and A will begin to decelerate as it slides against the friction on the table. •————————— Since the string is inextensible.

From ①, $-0.08g = 0.4a'$ •————————— Put $T = 0$ in equation ① as string is now slack.

$a' = -0.2g$ •————————————————— This is the new acceleration of A along the table.

$u_A = 2.42487, v = 0, a' = -0.2g, s = ?$

$v^2 = u^2 + 2as$

$0^2 = 2.42487^2 - 0.4gs$

$s = 1.5\,\text{m}$

A slides a further $1.5\,\text{m}$ along the table before it comes to rest.

∴ Total distance moved by A is $0.5 + 1.5 = 2.0\,\text{m}$.

Example 16

Two particles P and Q of masses 5 kg and 10 kg respectively are connected by a light inextensible string. The string passes over a small smooth pulley which is fixed at the top of a rough inclined plane. P rests on the inclined plane and Q hangs on the edge of the plane with the string vertical and taut. The plane is inclined to the horizontal at an angle α where $\tan \alpha = 0.75$, as shown in the diagram. The coefficient of friction between P and the plane is 0.2. The system is released from rest.

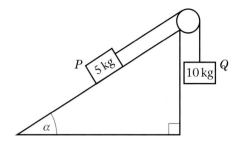

a Find the acceleration of the system.　　**b** Find the tension in the string.

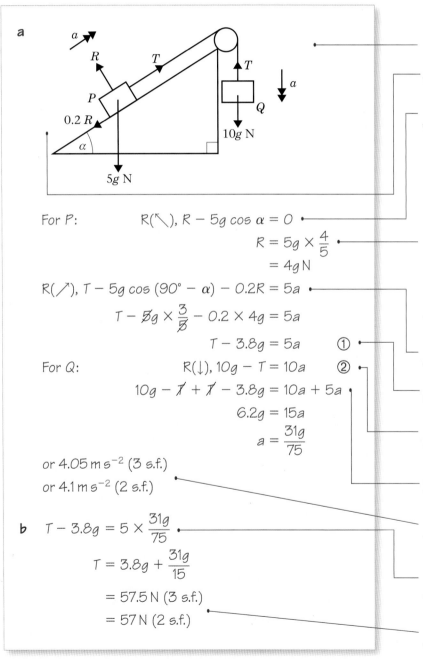

a

For P:　　　　$R(\nwarrow), R - 5g \cos \alpha = 0$

$$R = 5g \times \frac{4}{5}$$

$$= 4g \text{ N}$$

$R(\nearrow), T - 5g \cos (90° - \alpha) - 0.2R = 5a$

$$T - \cancel{5}g \times \frac{3}{\cancel{5}} - 0.2 \times 4g = 5a$$

$$T - 3.8g = 5a \qquad \text{①}$$

For Q:　　　　$R(\downarrow), 10g - T = 10a \qquad \text{②}$

$$10g - \cancel{T} + \cancel{T} - 3.8g = 10a + 5a$$

$$6.2g = 15a$$

$$a = \frac{31g}{75}$$

or 4.05 m s^{-2} (3 s.f.)

or 4.1 m s^{-2} (2 s.f.)

b　$T - 3.8g = 5 \times \dfrac{31g}{75}$

$$T = 3.8g + \frac{31g}{15}$$

$$= 57.5 \text{ N (3 s.f.)}$$

$$= 57 \text{ N (2 s.f.)}$$

Draw a diagram showing all the forces acting on each particle and their accelerations. Friction will be limiting.

No acceleration perpendicular to plane.

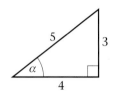

If　$\tan \alpha = \frac{3}{4}$,

　　$\cos \alpha = \frac{4}{5}$

and　$\sin \alpha = \frac{3}{5}$.

Resolve in the direction of the acceleration.

Substitute for R and simplify.

Resolve in the direction of the acceleration.

Add equations ① and ② to eliminate T.

Any of these answers would be acceptable.

Substitute for a in equation ①, using an unrounded value of a.

Either answer would be acceptable.

Exercise 3F

1 Two particles P and Q of mass 8 kg and 2 kg respectively, are connected by a light inextensible string. The particles are on a smooth horizontal plane. A horizontal force of magnitude F is applied to P in a direction away from Q and when the string is taut the particles move with acceleration $0.4\,\mathrm{m\,s^{-2}}$.

   a Find the value of F.

   b Find the tension in the string.

2 Two bricks P and Q, each of mass 5 kg, are connected by a light inextensible string. Brick P is held at rest and Q hangs freely, vertically below P. A force of 200 N is then applied vertically upwards to P causing it to accelerate at $1.2\,\mathrm{m\,s^{-2}}$. Assuming there is a resistance to the motion of each of the bricks of magnitude $R\,$N, find

   a the value of R,

   b the tension in the string connecting the bricks.

3 A car of mass 1500 kg is towing a trailer of mass 500 kg along a straight horizontal road. The car and the trailer are connected by a light inextensible tow-bar. The engine of the car exerts a driving force of magnitude 10 000 N and the car and the trailer experience resistances of magnitudes 3000 N and 1000 N respectively.

   a Find the acceleration of the car.

   b Find the tension in the tow-bar.

4 Two particles A and B of mass 4 kg and 3 kg respectively are connected by a light inextensible string which passes over a small smooth fixed pulley. The particles are released from rest with the string taut.

   a Find the tension in the string.

   When A has travelled a distance of 2 m it strikes the ground and immediately comes to rest.

   b Assuming that B does not hit the pulley find the greatest height that B reaches above its initial position.

5 Two particles A and B of mass 5 kg and 3 kg respectively are connected by a light inextensible string. Particle A lies on a rough horizontal table and the string passes over a small smooth pulley which is fixed at the edge of the table. Particle B hangs freely. The coefficient of friction between A and the table is 0.5. The system is released from rest. Find

   a the acceleration of the system,

   b the tension in the string,

   c the magnitude of the force exerted on the pulley by the string.

6 Two particles P and Q of equal mass are connected by a light inextensible string. The string passes over a small smooth pulley which is fixed at the top of a smooth inclined plane. The plane is inclined to the horizontal at an angle α where $\tan \alpha = 0.75$. Particle P is held at rest on the inclined plane at a distance of 2 m from the pulley and Q hangs freely on the edge of the plane at a distance of 3 m above the ground with the string vertical and taut. Particle P is released. Find the speed with which it hits the pulley.

7 Two particles P and Q of equal mass are connected by a light inextensible string. The string passes over a small smooth pulley which is fixed at the top of a fixed wedge. One face of the wedge is smooth and inclined to the horizontal at an angle of 30° and the other face of the wedge is rough and inclined to the horizontal at an angle of 60°. Particle P lies on the rough face and particle Q lies on the smooth face with the string connecting them taut. The coefficient of friction between P and the rough face is 0.5.

  a Find the acceleration of the system.

  b Find the tension in the string.

8 A van of mass 900 kg is towing a trailer of mass 500 kg up a straight road which is inclined to the horizontal at an angle α where $\tan \alpha = 0.75$. The van and the trailer are connected by a light inextensible tow-bar. The engine of the van exerts a driving force of magnitude 12 kN and the van and the trailer experience resistances to motion of magnitudes 1600 N and 600 N respectively.

  a Find the acceleration of the van.

  b Find the tension in the tow-bar.

9 Two particles P and Q of mass 2 kg and 3 kg respectively are connected by a light inextensible string. The string passes over a small smooth pulley which is fixed at the top of a rough inclined plane. The plane is inclined to the horizontal at an angle of 30°. Particle P is held at rest on the inclined plane and Q hangs freely on the edge of the plane with the string vertical and taut. Particle P is released and it accelerates up the plane at 2.5 m s^{-2}. Find

  a the tension in the string,

  b the coefficient of friction between P and the plane,

  c the force exerted by the string on the pulley.

10 A car of mass 900 kg is towing a trailer of mass 300 kg along a straight horizontal road. The car and the trailer are connected by a light inextensible tow-bar and when the speed of the car is 20 m s^{-1} the brakes are applied. This produces a braking force of 2400 N. Find

  a the deceleration of the car,

  b the magnitude of the force in the tow-bar,

  c the distance travelled by the car before it stops.

3.7 You can calculate the momentum of a particle and the impulse of a force.

■ **The momentum of a body of mass m which is moving with velocity v is mv.**

If m is in kg and v is in m s^{-1} then the momentum will be kg m s^{-1}. However, since kg m s^{-1} = (kg m s^{-2}) s and kg m s^{-2} are the units for force ($F = ma$) you can also measure momentum in N s.

> Velocity is a vector quantity and mass is a scalar, so momentum is a vector quantity.

Example 17

Find the magnitude of the momentum of

a a cricket ball of mass 400 g moving at $18 \, \text{m s}^{-1}$,

b a lorry of mass 5 tonnes moving at $0.3 \, \text{m s}^{-1}$.

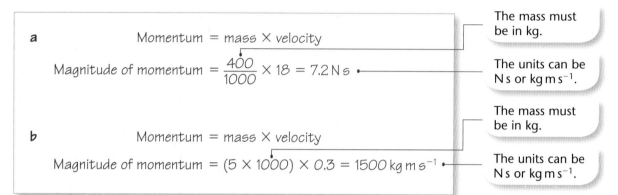

a   Momentum = mass × velocity

Magnitude of momentum = $\dfrac{400}{1000} \times 18 = 7.2 \, \text{N s}$

The mass must be in kg.

The units can be N s or kg m s^{-1}.

b   Momentum = mass × velocity

Magnitude of momentum = $(5 \times 1000) \times 0.3 = 1500 \, \text{kg m s}^{-1}$

The mass must be in kg.

The units can be N s or kg m s^{-1}.

■ **If a constant force _F_ acts for time _t_ then we define the impulse of the force to be _Ft_.**

If F is in N and t is in s then the units of impulse will be N s.

Force is a vector quantity and time is a scalar, so impulse is a vector quantity.

Examples of an impulse include a bat hitting a ball, a snooker ball hitting another ball or a jerk in a string when it suddenly goes tight. In all these cases the time for which the force acts is very small but the force is quite large and so the product of the two, which gives the impulse, is of reasonable size. However, there is no theoretical limit on the size of t.

Suppose a body of mass m is moving with an initial velocity u and is then acted upon by a force F for time t. This results in its final velocity being v.

Its acceleration is given by $a = \dfrac{v - u}{t}$.

Substituting into $F = ma$: $F = m\left(\dfrac{v - u}{t}\right)$

$$Ft = m(v - u)$$

$$= mv - mu$$

The impulse of the force I is given by: $I = Ft$.

■ $I = mv - mu$

Impulse = Final momentum − Initial momentum

Impulse = Change in momentum

This is a vector equation, so a positive direction must be chosen and each value given the correct sign.

This is called the **Impulse–Momentum Principle**.

Example 18

A body of mass 2 kg is initially at rest on a smooth horizontal plane. A horizontal force of magnitude 4.5 N acts on the body for 6 s. Find

a the magnitude of the impulse given to the body by the force,

b the final speed of the body.

a Magnitude of the impulse = force × time

$$= 4.5 \times 6 = 27 \,\text{N}\,\text{s}.$$ ———— The units can be N s or $kg\,m\,s^{-1}$.

b Impulse = Final momentum − Initial momentum

$$27 = 2v - 0$$ ———— The body is at rest initially.

$$v = 13.5 \,\text{m}\,\text{s}^{-1}$$

Example 19

A ball of mass 0.2 kg hits a fixed vertical wall at right angles with speed $3.5 \,\text{m}\,\text{s}^{-1}$. The ball rebounds with speed $2.5 \,\text{m}\,\text{s}^{-1}$. Find the magnitude of the impulse exerted on the wall by the ball.

This diagram shows the initial and final velocities of the ball and the impulse acting on it.

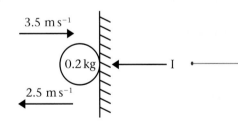

3.5 m s⁻¹

0.2 kg ←— I

2.5 m s⁻¹

$(\leftarrow) \ I = (0.2 \times 2.5) - (0.2 \times (-3.5))$

$= 0.5 + 0.7$

$= 1.2 \,\text{N}\,\text{s}$

Therefore, by Newton's 3rd Law, the magnitude of the impulse exerted on the wall by the ball is 1.2 N s.

Because the wall is fixed you cannot apply the Impulse–Momentum Principle to it. Find the magnitude of the impulse exerted on the ball by the wall and then use Newton's 3rd Law to deduce that the magnitude of the impulse exerted on the wall by the ball with be the same.

Note that this is a plan view of the situation.

Choose a positive direction and apply the Impulse–Momentum Principle to the ball.

The initial velocity is in the negative direction.

Exercise 3G

1 A ball of mass 0.5 kg is at rest when it is struck by a bat and receives an impulse of 15 N s. Find its speed immediately after it is struck.

2 A ball of mass 0.3 kg moving along a horizontal surface hits a fixed vertical wall at right angles with speed $3.5\,\mathrm{m\,s^{-1}}$. The ball rebounds at right angles to the wall. Given that the magnitude of the impulse exerted on the ball by the wall is 1.8 N s, find the speed of the ball just after it has hit the wall.

3 A ball of mass 0.2 kg is dropped from a height of 2.5 m above horizontal ground. After hitting the ground it rises to a height of 1.8 m above the ground. Find the magnitude of the impulse received by the ball from the ground.

4 A ball of mass 0.2 kg, moving along a horizontal surface, hits a fixed vertical wall at right angles. The ball rebounds at right angles to the wall with speed $3.5\,\mathrm{m\,s^{-1}}$. Given that the magnitude of the impulse exerted on the ball by the wall is 2 N s, find the speed of the ball just before it hit the wall.

5 A toy car of mass 0.2 kg is pushed from rest along a smooth horizontal floor by a horizontal force of magnitude 0.4 N for 1.5 s. Find its speed at the end of the 1.5 s.

3.8 You can solve problems involving collisions using the principle of Conservation of Momentum.

By Newton's 3rd Law, when two bodies collide each one exerts an equal and opposite force on the other. They are in contact for the same time, so they each exert an impulse on the other of equal magnitude but opposite in direction.

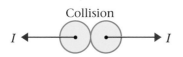

By the Impulse–Momentum Principle the changes in momentum of each body are equal but opposite in direction. Thus, these changes in momentum cancel each other out, and the momentum of the whole system is unchanged. This called the principle of **conservation of momentum**.

■ **Total momentum before impact = Total momentum after impact**

You can write this in symbols for two masses m_1 and m_2 with velocities u_1 and u_2 respectively before the collision, and velocities v_1 and v_2 respectively after the collision:

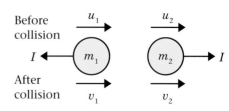

■ $m_1u_1 + m_2u_2 = m_1v_1 + m_2v_2$

When solving problems involving collisions, always:

- draw a diagram showing the velocities before and after the collision with arrows
- if appropriate, include the impulses on your diagram with arrows
- choose a positive direction and apply the Impulse–Momentum Principle and/or the principle of Conservation of Momentum.

Example 20

A particle P of mass 2 kg is moving with speed $3\,\text{m s}^{-1}$ on a smooth horizontal plane. Particle Q of mass 3 kg is at rest on the plane. Particle P collides with particle Q and after the collision Q moves off with speed $2\frac{1}{3}\,\text{m s}^{-1}$. Find

a the speed and direction of motion of P after the collision,

b the magnitude of the impulse received by P in the collision.

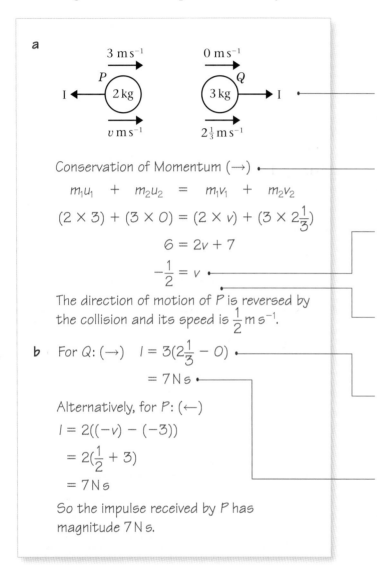

a

$3\,\text{m s}^{-1}$ $0\,\text{m s}^{-1}$

P 2 kg   3 kg Q → I

$v\,\text{m s}^{-1}$ $2\frac{1}{3}\,\text{m s}^{-1}$

Draw a diagram showing the velocities before and after the collision (with arrows) and the impulses (with arrows).

Conservation of Momentum $(\rightarrow)$

$$m_1u_1 \;+\; m_2u_2 \;=\; m_1v_1 \;+\; m_2v_2$$

$$(2 \times 3) + (3 \times 0) = (2 \times v) + (3 \times 2\tfrac{1}{3})$$

$$6 = 2v + 7$$

$$-\tfrac{1}{2} = v$$

Choose a positive direction and apply the principle of Conservation of Momentum.

Since v is negative, P must move to the left after the collision.

The direction of motion of P is reversed by the collision and its speed is $\frac{1}{2}\,\text{m s}^{-1}$.

The direction of motion of P in your answer must be with reference to the original direction of motion of P. Do not use the words left or right.

b For Q: $(\rightarrow)$ $I = 3(2\tfrac{1}{3} - 0)$

$$= 7\,\text{N s}$$

Alternatively, for P: $(\leftarrow)$

$$I = 2((-v) - (-3))$$

$$= 2(\tfrac{1}{2} + 3)$$

$$= 7\,\text{N s}$$

So the impulse received by P has magnitude $7\,\text{N s}$.

To find the impulse we must consider one particle and apply the Impulse–Momentum Principle. Here it is easier to consider Q.

Since each particle receives an impulse of equal magnitude, the magnitude of the impulse received by P is also $7\,\text{N s}$.

Example 21

Two particles A and B of mass 2 kg and 4 kg respectively are moving towards each other along the same straight line on a smooth horizontal surface. The particles collide. Before the collision the speeds of A and B are $3\,\text{m s}^{-1}$ and $2\,\text{m s}^{-1}$ respectively. After the collision the direction of motion of A is reversed and its speed is $2\,\text{m s}^{-1}$. Find

a the speed and direction of motion of B after the collision,

b the magnitude of the impulse given by A to B in the collision.

a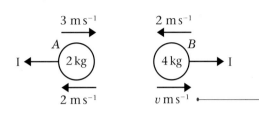

3 m s⁻¹ → A ← 2 m s⁻¹ → B
I ← (2 kg) (4 kg) → I
← 2 m s⁻¹ v m s⁻¹

'Guess' the direction of B after the collision. If it is moving in the opposite direction the answer will be negative.

This defines our positive direction.

Conservation of Momentum (→)

$(2 \times 3) + (4 \times -2) = (2 \times -2) + 4v$

$6 \quad - \quad 8 \quad = \quad -4 \quad + 4v$

$2 \quad = \quad 4v$

$0.5 \quad = \quad v$

Each velocity must be given the correct sign.

Since this value of v is positive we have guessed its direction after the collision correctly.

B has speed 0.5 m s⁻¹ and its direction of motion is reversed by the collision

b For A: (←):

Impulse–Momentum Principle

$I = 2(2 - -3)$

$= 10$ N s

The magnitude of the impulse given by A and B is 10 N s.

Although we could consider either particle it is safer to consider A since its initial and final speed were given in the question.

Choose a positive direction.

The magnitude of the impulse given by A to B is the same as the magnitude of the impulse given by B to A.

Example 22

Two particles P and Q, of masses 8 kg and 2 kg respectively, are connected by a light inextensible string. The particles are at rest on a smooth horizontal plane with the string slack. Particle P is projected directly away from Q with speed 4 m s⁻¹.

a Find the common speed of the particles when the string goes taut.

b Find the magnitude of the impulse transmitted through the string when it goes taut.

a

0 m s⁻¹ → 4 m s⁻¹ →
Q (2 kg) → I I ← (8 kg) P
v m s⁻¹ → v m s⁻¹ →

Draw a diagram showing all the speeds (with arrows) and the impulses (with arrows).

The string is inextensible so these are the same.

Using Conservation of Momentum (→):

$(2 \times 0) + (8 \times 4) = 2v + 8v$

$32 = 10v$

$3.2 = v$

This must be applied to the whole system.

The common speed of the particles is 3.2 m s⁻¹.

b For Q (→):

$I = 2(v - 0)$

$= 2 \times 3.2$

$= 6.4\,\text{N}\,\text{s}$

The magnitude of the impulse transmitted through the string (the 'jerk') is $6.4\,\text{N}\,\text{s}$.

To find the impulse we must consider one of the particles and apply the Impulse–Momentum Principle. It is easier to consider Q.

Example 23

Two particles A and B of mass 2 kg and 4 kg respectively are moving towards each other along the same straight line on a smooth horizontal surface. The particles collide. Before the collision the speeds of A and B are $3\,\text{m}\,\text{s}^{-1}$ and $2\,\text{m}\,\text{s}^{-1}$ respectively. Given that the magnitude of the impulse due to the collision is $7\,\text{N}\,\text{s}$, find

a the speed and direction of A after the collision,

b the speed and direction of B after the collision.

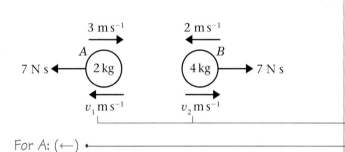

The diagram should show the velocities and impulses with arrows.

Again we must guess which way the particles go after the collision. There are 3 sensible possibilities – the one shown or $\overrightarrow{v_1}\ \overrightarrow{v_2}$ or $\overleftarrow{v_1}\ \overleftarrow{v_2}$

For A: (←)

$7 = 2(v_1 - -3)$

$7 = 2(v_1 + 3)$

$3.5 = v_1 + 3$

$0.5 = v_1$

To find either v_1 or v_2 we must consider one particle only, choose a positive direction and apply the Impulse–Momentum Principle.

For B: (→)

$7 = 4(v_2 - (-2))$

$1.75 = v_2 + 2$

$-0.25 = v_2$

$v_1 > 0$, so our guess for the direction of A was correct.

a The direction of motion of A is reversed and its speed is $0.5\,\text{m}\,\text{s}^{-1}$.

$v_2 < 0$, so our guess for the direction of B was incorrect.

b The direction of motion of B is unchanged and its speed is $0.25\,\text{m}\,\text{s}^{-1}$.

Note the form of the answers.

Exercise 3H

1 A particle P of mass 2 kg is moving on a smooth horizontal plane with speed $4 \, \text{m s}^{-1}$. It collides with a second particle Q of mass 1 kg which is at rest. After the collision P has speed $2 \, \text{m s}^{-1}$ and it continues to move in the same direction. Find the speed of Q after the collision.

2 A railway truck of mass 25 tonnes moving at $4 \, \text{m s}^{-1}$ collides with a stationary truck of mass 20 tonnes. As a result of the collision the trucks couple together. Find the common speed of the trucks after the collision.

3 Particles A and B have mass 0.5 kg and 0.2 kg respectively. They are moving with speeds $5 \, \text{m s}^{-1}$ and $2 \, \text{m s}^{-1}$ respectively in the same direction along the same straight line on a smooth horizontal surface when they collide. After the collision A continues to move in the same direction with speed $4 \, \text{m s}^{-1}$. Find the speed of B after the collision.

4 A particle of mass 2 kg is moving on a smooth horizontal plane with speed $4 \, \text{m s}^{-1}$. It collides with a second particle of mass 1 kg which is at rest. After the collision the particles join together.

   a Find the common speed of the particles after the collision.

   b Find the magnitude of the impulse in the collision.

5 Two particles A and B of mass 2 kg and 5 kg respectively are moving towards each other along the same straight line on a smooth horizontal surface. The particles collide. Before the collision the speeds of A and B are $6 \, \text{m s}^{-1}$ and $4 \, \text{m s}^{-1}$ respectively. After the collision the direction of motion of A is reversed and its speed is $1.5 \, \text{m s}^{-1}$. Find

   a the speed and direction of B after the collision,

   b the magnitude of the impulse given by A to B in the collision.

6 A particle P of mass 150 g is at rest on a smooth horizontal plane. A second particle Q of mass 100 g is projected along the plane with speed $u \, \text{m s}^{-1}$ and collides directly with P. On impact the particles join together and move on with speed $4 \, \text{m s}^{-1}$. Find the value of u.

7 A particle A of mass $4m$ is moving along a smooth horizontal surface with speed $2u$. It collides with another particle B of mass $3m$ which is moving with the same speed along the same straight line but in the opposite direction. Given that A is brought to rest by the collision, find

   a the speed of B after the collision and state its direction of motion,

   b the magnitude of the impulse given by A to B in the collision.

8 An explosive charge of mass 150 g is designed to split into two parts, one with mass 100 g and the other with mass 50 g. When the charge is moving at $4 \, \text{m s}^{-1}$ it splits and the larger part continues to move in the same direction whilst the smaller part moves in the opposite direction. Given that the speed of the larger part is twice the speed of the smaller part, find the speeds of the two parts.

9 Two particles P and Q of mass m and km respectively are moving towards each other along the same straight line on a smooth horizontal surface. The particles collide. Before the collision the speeds of P and Q are $3u$ and u respectively. After the collision the direction of motion of both particles is reversed and the speed of each particle is halved.

   a Find the value of k.

   b Find, in terms of m and u, the magnitude of the impulse given by P to Q in the collision.

10 Two particles A and B of mass 4 kg and 2 kg respectively are connected by a light inextensible string. The particles are at rest on a smooth horizontal plane with the string slack. Particle A is projected directly away from B with speed u m s^{-1}. When the string goes taut the impulse transmitted through the string has magnitude 6 N s. Find

   a the common speed of the particles just after the string goes taut,

   b the value of u.

11 Two particles P and Q of mass 3 kg and 2 kg respectively are moving along the same straight line on a smooth horizontal surface. The particles collide. After the collision both the particles are moving in the same direction, the speed of P is 1 m s^{-1} and the speed of Q is 1.5 m s^{-1}. The magnitude of the impulse of P on Q is 9 N s. Find

   a the speed and direction of P before the collision,

   b the speed and direction of Q before the collision.

12 Two particles A and B are moving in the same direction along the same straight line on a smooth horizontal surface. The particles collide. Before the collision the speed of B is 1.5 m s^{-1}. After the collision the direction of motion of both particles is unchanged, the speed of A is 2.5 m s^{-1} and the speed of B is 3 m s^{-1}. Given that the mass of A is three times the mass of B,

   a find the speed of A before the collision.

   Given that the magnitude of the impulse on A in the collision is 3 N s

   b find the mass of A.

Mixed exercise 3I

1 A bullet is fired by a gun which is 4 kg heavier than the bullet. Immediately after the bullet is fired, it is moving with speed 200 m s^{-1} and the gun recoils in the opposite direction with speed 5 m s^{-1}. Find

   a the mass of the bullet,

   b the mass of the gun.

2 A child of mass 25 kg moves from rest down a slide which is inclined to the horizontal at an angle of 35°. When the child has moved a distance of 4 m, her speed is 6 m s^{-1}. By modelling the child as a particle, find the coefficient of friction between the child and the slide.

3 A particle P of mass $3m$ is moving along a straight line with constant speed $2u$. It collides with another particle Q of mass $4m$ which is moving with speed u along the same line but in the opposite direction. As a result of the collision P is brought to rest.

a Find the speed of Q after the collision and state its direction of motion.

b Find the magnitude of the impulse exerted by Q on P in the collision.

4 A small box of mass $2\,\text{kg}$ is projected with speed $2\,\text{m}\,\text{s}^{-1}$ up a line of greatest slope of a rough plane. The plane is inclined to the horizontal at an angle α where $\tan\alpha = 0.75$. The coefficient of friction between the box and the plane is 0.4. The box is projected from the point P on the plane.

a Find the distance that the box travels up the plane before coming to rest.

b Show that the box will slide back down the plane.

c Find the speed of the box when it reached the point P.

5 Peter is pulling Paul, who is on a tobaggan, along a rough horizontal snow surface using a rope which makes an angle of $30°$ with the horizontal. Paul and the toboggan have a total mass of $40\,\text{kg}$ and the tobaggan is moving in a straight line with constant speed. The rope is modelled as a light inextensible string. Given that the tension in the rope is $50\,\text{N}$, find the coefficient of friction between the toboggan and the snow.

6 A particle of mass $0.5\,\text{kg}$ is pushed up a line of greatest slope of a rough plane by a horizontal force of magnitude $P\,\text{N}$. The plane is inclined to the horizontal at an angle α where $\tan\alpha = 0.75$ and the coefficient of friction between P and the plane is 0.5. The particle moves with constant speed. Find

a the magnitude of the normal reaction between the particle and the plane,

b the value of P.

7 A pile driver consists of a pile of mass $200\,\text{kg}$ which is knocked into the ground by dropping a driver of mass $1000\,\text{kg}$ onto it. The driver is released from rest at a point which is $10\,\text{m}$ vertically above the pile. Immediately after the driver impacts with the pile it can be assumed that they both move off with the same speed. By modelling the pile and the driver as particles,

a find the speed of the driver immediately before it hits the pile,

b find the common speed of the pile and driver immediately after the impact.

The ground provides a constant resistance to the motion of the pile driver of magnitude $120\,000\,\text{N}$.

c Find the distance that the pile driver is driven into the ground before coming to rest.

8 Particles P and Q of masses $2m$ and m respectively are attached to the ends of a light inextensible string which passes over a smooth fixed pulley. They both hang at a distance of $2\,\text{m}$ above horizontal ground. The system is released from rest.

a Find the magnitude of the acceleration of the system.

b Find the speed of P as it hits the ground.

Given that particle Q does not reach the pulley,

c find the greatest height that Q reaches above the ground.

d State how you have used in your calculation
   i   the fact that the string is inextensible,
   ii the fact that the pulley is smooth.

9 Two blocks A and B, of masses 20 kg and 30 kg respectively, are inside a cage of mass 10 kg. Block A is on top of block B. The blocks are lowered to the ground using a rope which is attached to the cage. The rope is modelled as a light inextensible string. Given that the blocks are moving vertically downwards with acceleration $0.8 \, \mathrm{m\,s^{-2}}$, find

a the tension in the rope,

b the magnitude of the force that block B exerts on block A,

c the normal reaction between block B and the floor of the cage.

10 A man, of mass 86 kg, is standing in a lift which is moving upwards with constant acceleration $2 \, \mathrm{m\,s^{-2}}$. Find the magnitude and direction of the force that the man is exerting on the floor of the lift.   **E**

11 A car, of mass 800 kg is travelling along a straight horizontal road. A constant retarding force of F N reduces the speed of the car from $18 \, \mathrm{m\,s^{-1}}$ to $12 \, \mathrm{m\,s^{-1}}$ in 2.4 s. Calculate

a the value of F,

b the distance moved by the car in these 2.4 s.   **E**

12 Two particles A and B, of mass 0.2 kg and 0.3 kg respectively, are free to move in a smooth horizontal groove. Initially B is at rest and A is moving toward B with a speed of $4 \, \mathrm{m\,s^{-1}}$. After the impact the speed of B is $1.5 \, \mathrm{m\,s^{-1}}$. Find

a the speed of A after the impact,

b the magnitude of the impulse of B on A during the impact.   **E**

13 A railway truck P of mass 2000 kg is moving along a straight horizontal track with speed $10 \, \mathrm{m\,s^{-1}}$. The truck P collides with a truck Q of mass 3000 kg, which is at rest on the same track. Immediately after the collision Q moves with speed $5 \, \mathrm{m\,s^{-1}}$. Calculate

a the speed of P immediately after the collision,

b the magnitude of the impulse exerted by P on Q during the collision.   **E**

14 A particle P of mass 1.5 kg is moving along a straight horizontal line with speed $3 \, \mathrm{m\,s^{-1}}$. Another particle Q of mass 2.5 kg is moving, in the opposite direction, along the same straight line with speed $4 \, \mathrm{m\,s^{-1}}$. The particles collide. Immediately after the collision the direction of motion of P is reversed and its speed is $2.5 \, \mathrm{m\,s^{-1}}$.

a Calculate the speed of Q immediately after the impact.

b State whether or not the direction of motion of Q is changed by the collision.

c Calculate the magnitude of the impulse exerted by Q on P, giving the units of your answer.   **E**

15 A particle A of mass m is moving with speed $2u$ in a straight line on a smooth horizontal table. It collides with another particle B of mass km which is moving in the same straight line on the table with speed u in the opposite direction to A. In the collision, the particles form a single particle which moves with speed $\frac{2}{3}u$ in the original direction of A's motion.

Find the value of k. **E**

16 A block of mass $0.8\,\text{kg}$ is pushed along a rough horizontal floor by a constant horizontal force of magnitude $7\,\text{N}$. The speed of the block increases from $2\,\text{m s}^{-1}$ to $4\,\text{m s}^{-1}$ in a distance of $4.8\,\text{m}$. Calculate

a the magnitude of the acceleration of the block,

b the coefficient of friction between the block and the floor. **E**

17 A particle is sliding with acceleration $3\,\text{m s}^{-2}$ down a line of greatest slope of a fixed plane. The plane is inclined at $40°$ to the horizontal.

Calculate the coefficient of friction between the particle and the plane. **E**

18 A pebble of mass $0.3\,\text{kg}$ slides in a straight line on the surface of a rough horizontal concrete path. Its initial speed is $12.6\,\text{m s}^{-1}$. The coefficient of friction between the pebble and the path is $\frac{3}{7}$.

a Find the frictional force retarding the pebble.

b Find the total distance covered by the pebble before it comes to rest. **E**

19 A particle moves down a line of greatest slope of a smooth plane inclined at an angle θ to the horizontal. The particle starts from rest and travels $3.5\,\text{m}$ in time $2\,\text{s}$. Find the value of $\sin\theta$. **E**

20 A man of mass $80\,\text{kg}$ stands in a lift. The lift has mass $60\,\text{kg}$ and is being raised vertically by a cable attached to the top of the lift. Given that the lift with the man inside is rising with a constant acceleration of $0.6\,\text{m s}^{-2}$, find, to two significant figures,

a the magnitude of the force exerted by the lift on the man,

b the magnitude of the force exerted by the cable on the lift.

The lift starts from rest and, $5\,\text{s}$ after starting to rise, the coupling between the cable and the lift suddenly snaps. There is an emergency cable attached to the lift but this only becomes taut when the lift is at the level of its initial position. After the coupling snaps, the lift moves freely under gravity until it is suddenly brought to rest in its initial position by the emergency cable. By modelling the lift with the man inside as a particle moving freely under gravity,

c find, to two significant figures, the magnitude of the impulse exerted by the emergency cable on the lift when it brings the lift to rest.

The model used in calculating the value required in part **c** ignores any effect of air resistance.

d State, with a reason, whether the answer obtained in **c** is higher or lower than the answer which would be obtained using a model which did incorporate the effect of air resistance. **E**

21 A box of mass 2 kg is pulled up a rough plane face by means of a light rope. The plane is inclined at an angle of 20° to the horizontal, as shown in the figure. The rope is parallel to a line of greatest slope of the plane. The tension in the rope is 18 N. The coefficient of friction between the box and the plane is 0.6. By modelling the box as a particle, find

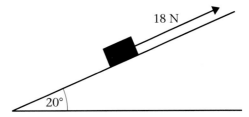

a the normal reaction of the plane on the box,

b the acceleration of the box.

22 A sledge has mass 30 kg. The sledge is pulled in a straight line along horizontal ground by means of a rope. The rope makes an angle 20° with the horizontal, as shown in the figure. The coefficient of friction between the sledge and the ground is 0.2. The sledge is modelled as a particle and the rope as a light inextensible string. The tension in the rope is 150 N. Find, to three significant figures,

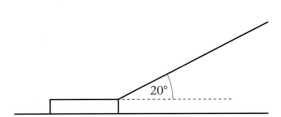

a the normal reaction of the ground on the sledge,

b the acceleration of the sledge.

When the sledge is moving at $12\,\text{m}\,\text{s}^{-1}$, the rope is released from the sledge.

c Find, to three significant figures, the distance travelled by the sledge from the moment when the rope is released to the moment when the sledge comes to rest.

23 A metal stake of mass 2 kg is driven vertically into the ground by a blow from a sledgehammer of mass 10 kg. The sledgehammer falls vertically on to the stake, its speed just before impact being $9\,\text{m}\,\text{s}^{-1}$. In a model of the situation it is assumed that, after impact, the stake and the sledgehammer stay in contact and move together before coming to rest.

a Find the speed of the stake immediately after impact.

The stake moves 3 cm into the ground before coming to rest. Assuming in this model that the ground exerts a constant resistive force of magnitude R newtons as the stake is driven down,

b find the value of R.

c State one way in which this model might be refined to be more realistic.

24 The particles have mass 3 kg and m kg, where $m < 3$. They are attached to the ends of a light inextensible string. The string passes over a smooth fixed pulley. The particles are held in position with the string taut and the hanging parts of the string vertical, as shown in the figure. The particles are then released from rest. The initial acceleration of each particle has magnitude $\frac{3}{7}g$. Find

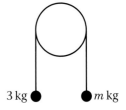

a the tension in the string immediately after the particles are released,

b the value of m.

25

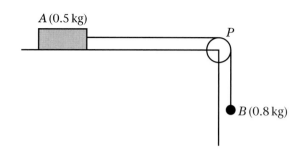

A block of wood A of mass 0.5 kg rests on a rough horizontal table and is attached to one end of a light inextensible string. The string passes over a small smooth pulley P fixed at the edge of the table. The other end of the string is attached to a ball B of mass 0.8 kg which hangs freely below the pulley, as shown in the figure. The coefficient of friction between A and the table is μ. The system is released from rest with the string taut. After release, B descends a distance of 0.4 m in 0.5 s. Modelling A and B as particles, calculate

a the acceleration of B,

b the tension in the string,

c the value of μ.

d State how in your calculations you have used the information that the string is inextensible.

E

26

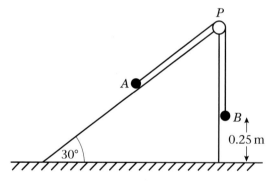

Two particles A and B, of mass m kg and 3 kg respectively, are connected by a light inextensible string. The particle A is held resting on a smooth fixed plane inclined at 30° to the horizontal. The string passes over a smooth pulley P fixed at the top of the plane. The portion AP of the string lies along a line of greatest slope of the plane and B hangs freely from the pulley, as shown in the figure. The system is released from rest with B at a height of 0.25 m above horizontal ground. Immediately after release, B descends with an acceleration of $\frac{2}{5}g$. Given that A does not reach P, calculate

a the tension in the string while B is descending,

b the value of m.

The particle B strikes the ground and does not rebound. Find

c the magnitude of the impulse exerted by B on the ground,

d the time between the instant when B strikes the ground and the instant when A reaches its highest point.

E

Summary of key points

1   The unit of force is the Newton (N). It is defined as the force that will cause a mass of 1 kg to accelerate at a rate of $1\,\text{m}\,\text{s}^{-2}$.

$$F = ma$$

2   The force due to gravity is called the weight of an object, and it acts vertically downwards. A particle falling freely experiences an acceleration of $g = 9.8\,\text{m}\,\text{s}^{-2}$.

$$W = mg$$

3   The component of a force of magnitude F acting in a certain direction is $F\cos\theta$, where θ is the size of the angle between the force and the direction.

4   The maximum or limiting value of the friction F_{MAX} between two surfaces is given by

$$F_{\text{MAX}} = \mu R$$

where μ is the coefficient of friction and R is the normal reaction between the two surfaces.

5   If a force P is applied to a block of mass m which is at rest on a rough horizontal surface and P acts at an angle to the horizontal:

- the normal reaction R is not equal to mg
- the force tending to pull or push the block along the plane is not equal to P.

6   A particle placed on a rough inclined plane will remain at rest if $\tan\theta \leqslant \mu$ where θ is the angle the plane makes with the horizontal and μ is the coefficient of friction between the particle and the plane.

7   Provided all parts of a connected system are moving in the same straight line you can treat the whole system as a single particle.

8   In problems involving particles which are connected by string(s) which pass over pulley(s) you cannot treat the whole system as a single particle. This is because the particles are moving in different directions.

9   The momentum of a body of mass m which is moving with velocity v is mv.

10 If a constant force F acts for time t then we define the impulse of the force to be Ft.

11 The Impulse–Momentum Principle states that:

$$\text{Impulse} = \text{Final momentum} - \text{Initial momentum} = \text{Change in momentum}$$
$$I \qquad = \qquad mv \qquad - \qquad mu$$

12 The principle of Conservation of Momentum states that:

$$\frac{\text{Total momentum}}{\text{before impact}} = \frac{\text{Total momentum}}{\text{after impact}}$$

$$m_1 u_1 + m_2 u_2 = m_1 v_1 + m_2 v_2$$

1

Review Exercise

1 A train decelerates uniformly from $35 \, \text{m s}^{-1}$ to $21 \, \text{m s}^{-1}$ in a distance of 350 m. Calculate

  a the deceleration,

  b the total time taken, under this deceleration, to come to rest from a speed of $35 \, \text{m s}^{-1}$.   **E**

2 In taking off, an aircraft moves on a straight runway AB of length 1.2 km. The aircraft moves from A with initial speed $2 \, \text{m s}^{-1}$. It moves with constant acceleration and 20s later it leaves the runway at C with speed $74 \, \text{m s}^{-1}$. Find

  a the acceleration of the aircraft,

  b the distance BC.   **E**

3 A car is moving along a straight horizontal road at a constant speed of $18 \, \text{m s}^{-1}$. At the instant when the car passes a lay-by, a motorcyclist leaves the lay-by, starting from rest, and moves with constant acceleration $2.5 \, \text{m s}^{-2}$ in pursuit of the car. Given that the motorcyclist overtakes the car T seconds after leaving the lay-by, calculate

  a the value of T,

  b the speed of the motorcyclist at the instant of passing the car.   **E**

4 A particle moves with constant acceleration along the straight line OLM and passes through the points O, L and M at times 0 s, 4 s and 10 s respectively. Given that $OL = 14$ m and $OM = 50$ m, find

  a the acceleration of the particle,

  b the speed of the particle at M.   **E**

5 A particle P moves in a straight line with constant retardation. At the instants when P passes through the points A, B and C, it is moving with speeds $10 \, \text{m s}^{-1}$, $7 \, \text{m s}^{-1}$ and $3 \, \text{m s}^{-1}$ respectively.

  Show that $\dfrac{AB}{BC} = \dfrac{51}{40}$.   **E**

6 A car moving with uniform acceleration along a straight level road, passed points A and B when moving with speed $6 \, \text{m s}^{-1}$ and $14 \, \text{m s}^{-1}$ respectively. Find the speed of the car at the instant that it passed C, the mid-point of AB.   **E**

7 A particle P is moving along the x-axis with constant deceleration $3\,\text{m}\,\text{s}^{-2}$. At time $t = 0\,\text{s}$, P is passing through the origin O and is moving with speed $u\,\text{m}\,\text{s}^{-1}$ in the direction of x increasing. At time $t = 8\,\text{s}$, P is instantaneously at rest at the point A. Find

  a the value of u,

  b the distance OA,

  c the times at which P is $24\,\text{m}$ from A.

8 A train moves along a straight track with constant acceleration. Three telegraph poles are set at equal intervals beside the track at points A, B and C, where $AB = 50\,\text{m}$ and $BC = 50\,\text{m}$. The front of the train passes A with speed $22.5\,\text{m}\,\text{s}^{-1}$, and $2\,\text{s}$ later it passes B. Find

  a the acceleration of the train,

  b the speed of the front of the train when it passes C,

  c the time that elapses from the instant the front of the train passes B to the instant it passes C. **E**

9 A particle X, moving along a straight line with constant speed $4\,\text{m}\,\text{s}^{-1}$, passes through a fixed point O. Two seconds later, another particle Y, moving along the same straight line and in the same direction, passes through O with speed $6\,\text{m}\,\text{s}^{-1}$. The particle Y is moving with constant deceleration $2\,\text{m}\,\text{s}^{-2}$.

  a Write down expressions for the velocity and displacement of each particle t seconds after Y passed through O.

  b Find the shortest distance between the particles after they have both passed through O.

  c Find the value of t when the distance between the particles has increased to $23\,\text{m}$. **E**

10 A stone is projected vertically upwards from a point A with initial speed $u\,\text{m}\,\text{s}^{-1}$. It takes $3.5\,\text{s}$ to reach its maximum height above A. Find

  a the value of u,

  b the maximum height of the stone above A.

11 A small ball is projected vertically upwards from a point A. The greatest height reached by the ball is $40\,\text{m}$ above A. Calculate

  a the speed of projection,

  b the time between the instant that the ball is projected and the instant it returns to A. **E**

12 A ball is projected vertically upwards and takes 3 seconds to reach its highest point. At time t seconds, the ball is $39.2\,\text{m}$ above its point of projection. Find the possible values of t. **E**

13 A stone is thrown vertically upwards with speed $16\,\text{m}\,\text{s}^{-1}$ from a point h metres above the ground. The stone hits the ground $4\,\text{s}$ later. Find

  a the value of h,

  b the speed of the stone as it hits the ground. **E**

14 Two balls are projected simultaneously from two points A and B. The point A is $54\,\text{m}$ vertically below B. Initially one ball is projected from A towards B with speed $15\,\text{m}\,\text{s}^{-1}$. At the same time the other ball is projected from B towards A with speed $12\,\text{m}\,\text{s}^{-1}$.

Find the distance between A and the point where the balls collide.

15 A particle is projected vertically upwards from a point A with speed $u\,\text{m s}^{-1}$. The particle takes $2\frac{6}{7}\,\text{s}$ to reach its greatest height above A. Find

a the value of u,

b the total time for which the particle is more than $17\frac{1}{2}\,\text{m}$ above A.

16 A particle moves along a horizontal straight line. The particle starts from rest, accelerates at $2\,\text{m s}^{-2}$ for 3 seconds, and then decelerates at a constant rate coming to rest in a further 8 seconds.

a Sketch a speed–time graph to illustrate the motion of the particle.

b Find the total distance travelled by the particle during these 11 seconds. **E**

17 A man is driving a car on a straight horizontal road. He sees a junction S ahead, at which he must stop. When the car is at the point P, 300 m from S, its speed is $30\,\text{m s}^{-1}$. The car continues at this constant speed for 2 s after passing P. The man then applies the brakes so that the car has constant deceleration and comes to rest at S.

a Sketch a speed–time graph to illustrate the motion of the car in moving from P to S.

b Find the time taken by the car to travel from P to S. **E**

18

$v(\text{m s}^{-1})$

The figure shows the speed–time graph of a cyclist moving on a straight road over a 7 s period. The sections of the graph from $t = 0$ to $t = 3$, and from $t = 3$ to $t = 7$, are straight lines. The section from $t = 3$ to $t = 7$ is parallel to the t-axis.

State what can be deduced about the motion of the cyclist from the fact that

a the graph from $t = 0$ to $t = 3$ is a straight line,

b the graph from $t = 3$ to $t = 7$ is parallel to the t-axis.

c Find the distance travelled by the cyclist during this 7 s period. **E**

19 A train stops at two stations 7.5 km apart. Between the stations it takes 75 s to accelerate uniformly to a speed $24\,\text{m s}^{-1}$, then travels at this speed for a time T seconds before decelerating uniformly for the final 0.6 km.

a Draw a speed–time graph to illustrate this journey.

Hence, or otherwise, find

b the deceleration of the train during the final 0.6 km,

c the value of T,

d the total time for the journey. **E**

20 A car accelerates uniformly from rest to a speed of $20\,\text{m s}^{-1}$ in T seconds. The car then travels at a constant speed of $20\,\text{m s}^{-1}$ for $4T$ seconds and finally decelerates uniformly to rest in a further 50 s.

a Sketch a speed–time graph to show the motion of the car.

The total distance travelled by the car is 1220 m. Find

b the value of T,

c the initial acceleration of the car. **E**

21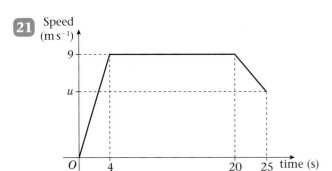

A sprinter runs a race of 200 m. Her total time for running the race is 25 s. The figure is a sketch of the speed–time graph for the motion of the sprinter. She starts from rest and accelerates uniformly to a speed of $9\,\mathrm{m\,s^{-1}}$ in 4 s. The speed of $9\,\mathrm{m\,s^{-1}}$ is maintained for 16 s and she then decelerates uniformly to a speed of $u\,\mathrm{m\,s^{-1}}$ at the end of the race. Calculate

a the distance covered by the sprinter in the first 20 s of the race,

b the value of u,

c the deceleration of the sprinter in the last 5 s of the race.   **E**

22 An electric train starts from rest at a station A and moves along a straight level track. The train accelerates uniformly at $0.4\,\mathrm{m\,s^{-2}}$ to a speed of $16\,\mathrm{m\,s^{-1}}$. The speed is then maintained for a distance of 2000 m. Finally the train retards uniformly for 20 s before coming to rest at a station B. For this journey from A to B,

a find the total time taken,

b find the distance from A to B,

c sketch the *distance*–time graph, showing clearly the shape of the graph for each stage of the journey.   **E**

23 A car starts from rest at a point S on straight racetrack. The car moves with constant acceleration for 20 s, reaching a speed of $25\,\mathrm{m\,s^{-1}}$. The car then travels at a constant speed of $25\,\mathrm{m\,s^{-1}}$ for 120 s. Finally it moves with constant deceleration, coming to rest at a point F.

a Sketch a speed–time graph to illustrate the motion of the car.

The distance between S and F is 4 km.

b Calculate the total time the car takes to travel from S to F.

A motorcycle starts at S, 10 s after the car has left S. The motorcycle moves with constant acceleration from rest and passes the car at a point P which is 1.5 km from S. When the motorcycle passes the car, the motorcycle is still accelerating and the car is moving at a constant speed. Calculate

c the time the motorcycle takes to travel from S to P,

d the speed of the motorcycle at P.   **E**

24 Two cars A and B are travelling in the same direction along a motorway. They pass a warning sign at the same instant and, subsequently, arrive at a toll booth at the same instant.

Car A passes the warning sign at speed $24\,\mathrm{m\,s^{-1}}$, continues at this speed for one minute, then decelerates uniformly, coming to rest at the toll booth.

Car B passes the warning sign at speed $30\,\mathrm{m\,s^{-1}}$, continues at this speed for T seconds, then decelerates uniformly, coming to rest at the toll booth.

a On the same diagram, draw a speed–time graph to illustrate the motion of each car.

The distance between the warning sign and the toll booth is 1.56 km.

b Find the length of time for which A is decelerating.

c Find the value of T.   **E**

25 A train is travelling at $10\,\mathrm{m\,s^{-1}}$ on a straight horizontal track. The driver sees a red signal 135 m ahead and immediately applies the brakes. The train immediately decelerates with constant deceleration for 12 s, reducing its speed to $3\,\mathrm{m\,s^{-1}}$. The driver then releases the brakes and allows the train to travel at a constant speed of $3\,\mathrm{m\,s^{-1}}$ for a further $15\,\mathrm{s}$. He then applies the brakes again and the train slows down with constant deceleration, coming to rest as it reaches the signal.

a Sketch a speed–time graph illustrating the motion of the train.

b Find the distance travelled by the train from the moment when the brakes are first applied to the moment when its speed first reaches $3\,\mathrm{m\,s^{-1}}$.

c Find the total time from the moment when the brakes are first applied to the moment when the train comes to rest.

d Sketch an acceleration–time graph illustrating the motion of the train. **E**

26 A straight stretch of railway line passes over a viaduct which is $600\,\mathrm{m}$ long. An express train on this stretch of line normally travels at a speed of $50\,\mathrm{m\,s^{-1}}$. Some structural weakness in the viaduct is detected and engineers specify that all trains passing over the viaduct must do so at a speed of no more than $10\,\mathrm{m\,s^{-1}}$. Approaching the viaduct, the train therefore reduces its speed from $50\,\mathrm{m\,s^{-1}}$ with constant deceleration $0.5\,\mathrm{m\,s^{-2}}$, reaching a speed of precisely $10\,\mathrm{m\,s^{-1}}$ just as it reaches the viaduct. It then passes over the viaduct at a constant speed of $10\,\mathrm{m\,s^{-1}}$. As soon as it reaches the other side, it accelerates to its normal speed of $50\,\mathrm{m\,s^{-1}}$ with constant acceleration $0.5\,\mathrm{m\,s^{-2}}$.

a Sketch a speed–time graph to show the motion of the train during the period from the time when it starts to reduce speed to the time when it is running at full speed again.

b Find the total distance travelled by the train while its speed is less than $50\,\mathrm{m\,s^{-1}}$.

c Find the extra time taken by the train for the journey due to the speed restriction on the viaduct. **E**

27 A bus and a cyclist are moving along a straight horizontal road in the same direction. The bus starts at a bus stop O and moves with constant acceleration of $2\,\mathrm{m\,s^{-2}}$ until it reaches a maximum speed of $12\,\mathrm{m\,s^{-1}}$. It then maintains this constant speed. The cyclist travels with a constant speed of $8\,\mathrm{m\,s^{-1}}$. The cyclist passes O just as the bus starts to move. The bus later overtakes the cyclist at the point A.

a Show that the bus does not overtake the cyclist before it reaches its maximum speed.

b Sketch, on the same diagram, speed–time graphs to represent the motion of the bus and the cyclist as they move from O to A.

c Find the time taken for the bus and the cyclist to move from O to A.

d Find the distance OA. **E**

28 Two trains A and B run on parallel straight tracks. Initially both are at rest in a station and level with each other. At time $t = 0$, A starts to move. It moves with constant acceleration for $12\,\mathrm{s}$ up to a speed of $30\,\mathrm{m\,s^{-1}}$, and then moves at a constant speed of $30\,\mathrm{m\,s^{-1}}$. Train B starts to move in the same direction as A when $t = 40$, where t is measured in seconds. It accelerates with the same initial acceleration as A, up to a speed of $60\,\mathrm{m\,s^{-1}}$. It then moves at a constant speed of $60\,\mathrm{m\,s^{-1}}$. Train B overtakes A after both trains have reached their maximum speed. Train B overtakes A when $t = T$.

a Sketch, on the same diagram, the speed–time graphs of both trains for $0 \leqslant t \leqslant T$.

b Find the value of T. **E**

29 A train starts from rest at a station, accelerates uniformly to its maximum speed of $15\,\text{m s}^{-1}$, travels at this speed for a time, and then decelerates uniformly to rest at the next station. The distance from station to station is $1260\,\text{m}$, and the time spent travelling at the maximum speed is three-quarters of the total journey time.

a Sketch a speed–time graph to illustrate this information.

b Find the total journey time.

Given also that the magnitude of the deceleration is twice the magnitude of the acceleration,

c find the magnitude of the acceleration. **E**

30 The brakes of a train, which is travelling at $108\,\text{km h}^{-1}$, are applied as the train passes a point A. The brakes produce a retardation of magnitude $3f\,\text{m s}^{-2}$ until the speed of the train is reduced to $36\,\text{km h}^{-1}$. The train travels at this speed for a distance and is then uniformly accelerated at $f\,\text{m s}^{-2}$ until it again reaches the speed $108\,\text{km h}^{-1}$ as it passes point B. The time taken by the train in travelling from A to B, a distance of $4\,\text{km}$, is 4 minutes.

a Sketch a speed–time graph to illustrate the motion of the train from A to B.

b Find the value of f.

c Find the distance travelled at $36\,\text{km h}^{-1}$. **E**

31 A car of mass $750\,\text{kg}$, moving along a level straight road, has its speed reduced from $25\,\text{m s}^{-1}$ to $15\,\text{m s}^{-1}$ by brakes which produce a constant retarding force of $2250\,\text{N}$. Calculate the distance travelled by the car as its speed is reduced from $25\,\text{m s}^{-1}$ to $15\,\text{m s}^{-1}$. **E**

32 A particle P is moving with constant acceleration along a straight horizontal line ABC, where $AC = 24\,\text{m}$. Initially P is at A and is moving with speed $5\,\text{m s}^{-1}$ in the direction AB. After $1.5\,\text{s}$, the direction of motion of P is unchanged and P is at B with speed $9.5\,\text{m s}^{-1}$.

a Show that the speed of P at C is $13\,\text{m s}^{-1}$.

The mass of P is $2\,\text{kg}$. When P reaches C, an impulse of magnitude $30\,\text{N s}$ is applied to P in the direction CB.

b Find the velocity of P immediately after the impulse has been applied, stating clearly the direction of motion of P at this instant. **E**

33 A ball of mass $0.3\,\text{kg}$ is released at rest from a point at a height of $10\,\text{m}$ above horizontal ground. After hitting the ground the ball rebounds to a height of $2.5\,\text{m}$.

Calculate the magnitude of the impulse exerted by the ground on the ball. **E**

34 Two particles A and B have mass $0.4\,\text{kg}$ and $0.3\,\text{kg}$ respectively. They are moving in opposite directions on a smooth horizontal table and collide directly. Immediately before the collision, the speed of A is $6\,\text{m s}^{-1}$ and the speed of B is $2\,\text{m s}^{-1}$. As a result of the collision, the direction of motion of B is reversed and its speed immediately after the collision is $3\,\text{m s}^{-1}$. Find

a the speed of A immediately after the collision, stating clearly whether the direction of motion of A is changed by the collision,

b the magnitude of the impulse exerted on B in the collision, stating clearly the units in which your answer is given. **E**

35 A railway truck S of mass 2000 kg is travelling due east along a straight horizontal track with constant speed $12\,\mathrm{m\,s}^{-1}$. The truck S collides with a truck T which is travelling due west along the same track as S with constant speed $6\,\mathrm{m\,s}^{-1}$. The magnitude of the impulse of T on S is 28 800 N s.

a Calculate the speed of S immediately after the collision.

b State the direction of motion of S immediately after the collision.

Given that, immediately after the collision, the speed of T is $3.6\,\mathrm{m\,s}^{-1}$, and that T and S are moving in opposite directions,

c calculate the mass of T. **E**

36 Two particles A and B, of mass 0.5 kg and 0.4 kg respectively, are travelling in the same straight line on a smooth horizontal table. Particle A, moving with speed $3\,\mathrm{m\,s}^{-1}$, strikes particle B, which is moving with speed $2\,\mathrm{m\,s}^{-1}$ in the same direction. After the collision A and B are moving in the same direction and the speed of B is $0.8\,\mathrm{m\,s}^{-1}$ greater than the speed of A.

a Find the speed of A and the speed of B after the collision.

b Show that A loses momentum 0.4 N s in the collision.

Particle B later hits an obstacle on the table and rebounds in the opposite direction with speed $1\,\mathrm{m\,s}^{-1}$.

c Find the magnitude of the impulse received by B in this second impact. **E**

37 Two particles A and B, of mass 3 kg and 2 kg respectively, are moving in the same direction on a smooth horizontal table when they collide directly. Immediately before the collision, the speed of A is $4\,\mathrm{m\,s}^{-1}$ and the speed of B is $1.5\,\mathrm{m\,s}^{-1}$. In the collision, the particles join to form a single particle C.

a Find the speed of C immediately after the collision.

Two particles P and Q have mass 3 kg and m kg respectively. They are moving towards each other in opposite directions on a smooth horizontal table. Each particle has speed $4\,\mathrm{m\,s}^{-1}$, when they collide directly. In this collision, the direction of motion of each particle is reversed. The speed of P immediately after the collision is $2\,\mathrm{m\,s}^{-1}$ and the speed of Q is $1\,\mathrm{m\,s}^{-1}$.

b Find

i the value of m,

ii the magnitude of the impulse exerted on Q in the collision. **E**

38 A railway truck, of mass 1500 kg and travelling with a speed $6\,\mathrm{m\,s}^{-1}$ along a horizontal track, collides with a stationary truck of mass 2000 kg. After the collision the two trucks move on together, coming to rest after 12 second. Calculate the magnitude of the constant force resisting their motion after the collision. **E**

39 A particle P of mass 0.3 kg is moving in a straight line on a rough horizontal plane. The speed of P decreases from $7.5\,\mathrm{m\,s}^{-1}$ to $4\,\mathrm{m\,s}^{-1}$ in time T seconds. Given the coefficient of friction between P and the plane is $\frac{1}{7}$, find

a the magnitude of the frictional force opposing the motion of P,

b the value of T. **E**

40 A small stone moves horizontally in a straight line across the surface of an ice rink. The stone is given an initial speed of $7\,\mathrm{m\,s}^{-1}$. It comes to rest after moving a distance of 10 m. Find

a the deceleration of the stone while it is moving,

b the coefficient of friction between the stone and the ice. **E**

41 A rough plane is inclined at an angle α to the horizontal, where $\tan\alpha = \frac{3}{4}$. A particle slides with acceleration $3.5\,\mathrm{m\,s^{-2}}$ down a line of greatest slope of this plane. Calculate the coefficient of friction between the particle and the plane. **E**

42 A particle moves down a line of greatest slope of a rough plane which is inclined at $30°$ to the horizontal. The particle starts from rest and moves $3.5\,\mathrm{m}$ in time $2\,\mathrm{s}$. Find the coefficient of friction between the particle and the plane. **E**

43 A stone S is sliding on ice. The stone is moving along a straight line ABC, where $AB = 24\,\mathrm{m}$ and $AC = 30\,\mathrm{m}$. The stone is subject to a constant resistance to motion of magnitude $0.3\,\mathrm{N}$. At A the speed of S is $20\,\mathrm{m\,s^{-1}}$, and at B the speed of S is $16\,\mathrm{m\,s^{-1}}$. Calculate

a the deceleration of S,

b the speed of S at C.

c Show that the mass of S is $0.1\,\mathrm{kg}$.

At C, the stone S hits a vertical wall, rebounds from the wall and then slides back along the line CA. The magnitude of the impulse of the wall on S is $2.4\,\mathrm{N\,s}$ and the stone continues to move against a constant resistance of $0.3\,\mathrm{N}$.

d Calculate the time between the instant that S rebounds from the wall and the instant that S comes to rest. **E**

44 A railway truck P of mass $1500\,\mathrm{kg}$ is moving on a straight horizontal track. The truck P collides with a truck Q of $2500\,\mathrm{kg}$ at a point A. Immediately before the collision, P and Q are moving in the same direction with speeds $10\,\mathrm{m\,s^{-1}}$ and $5\,\mathrm{m\,s^{-1}}$ respectively. Immediately after the collision, the direction of motion of P is unchanged and its speed is $4\,\mathrm{m\,s^{-1}}$. By modelling the trucks as particles,

a show that the speed of Q immediately after the collision is $8.6\,\mathrm{m\,s^{-1}}$.

After the collision at A, the truck P is acted upon by a constant braking force of magnitude $500\,\mathrm{N}$. The truck P comes to rest at the point B.

b Find the distance AB.

After the collision Q continues to move with constant speed $8.6\,\mathrm{m\,s^{-1}}$.

c Find the distance between P and Q at the instant when P comes to rest. **E**

45

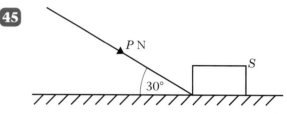

A heavy suitcase S of mass $50\,\mathrm{kg}$ is moving along a horizontal floor under the action of a force of magnitude P newtons. The force acts at $30°$ to the floor, as shown in the figure, and S moves in a straight line at constant speed. The suitcase is modelled as a particle and the floor as a rough horizontal plane. The coefficient of friction between S and the floor is $\frac{3}{5}$.

Calculate the value of P. **E**

46 An engine of mass 25 tonnes pulls a truck of mass 10 tonnes along a railway line. The frictional resistances to the motion of the engine and the truck are modelled as constant and of magnitude $50\,\mathrm{N}$ per tonne. When the train is travelling horizontally the tractive force exerted by the engine is $26\,\mathrm{kN}$. Modelling the engine and the truck as particles and the coupling between the engine and the truck as a light horizontal rod, calculate

a the acceleration of the engine and the truck,

b the tension in the coupling.

The engine and the truck now climb a slope which is modelled as a plane inclined at angle α to the horizontal, where $\sin \alpha = \frac{1}{70}$. The engine and the truck are moving up the slope with an acceleration of magnitude $0.6 \, \text{m s}^{-2}$. The frictional resistances to motion are modelled as before.

c Calculate the tractive force exerted by the engine. Give your answer in kN. (1 tonne = 1000 kg) **E**

47

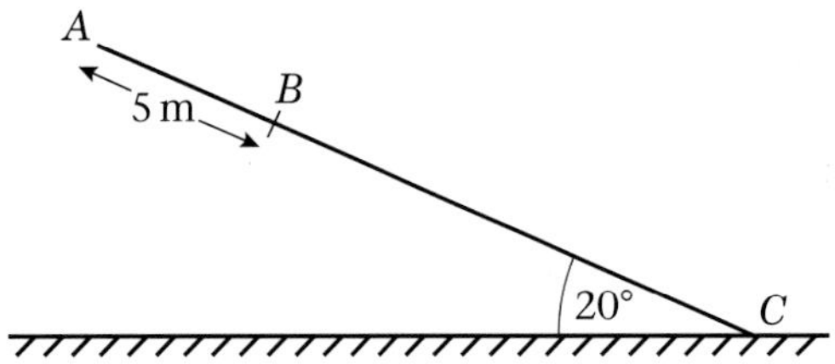

A suitcase of mass 10 kg slides down a ramp which is inclined at an angle of 20° to the horizontal. The suitcase is modelled as a particle and the ramp as a rough plane. The top of the plane is A. The bottom of the plane is C and AC is a line of greatest slope, as shown in the figure above. The point B is on AC with $AB = 5$ m. The suitcase leaves A with a speed of $10 \, \text{m s}^{-1}$ and passes B with a speed of $8 \, \text{m s}^{-1}$. Find

a the deceleration of the suitcase,

b the coefficient of friction between the suitcase and the ramp.

The suitcase reaches the bottom of the ramp.

c Find the greatest possible length of AC. **E**

48 A slipway for launching boats consists of a rough straight track inclined at an angle of 10° to the horizontal. A boat of mass 300 kg is pulled down the slipway by means of a rope which is parallel to the slipway. When the tension in the rope is 500 N, the boat moves down the slipway with constant speed.

a Find, to two significant figures, the coefficient of friction between the boat and the slipway.

Later the boat returns to the slipway. It is now pulled up the slipway at constant speed by the rope which is again parallel to the slipway.

b Give a brief reason why the magnitude of the frictional force is the same as when the boat is pulled down the slope.

c Find, to two significant figures, the tension in the rope. **E**

49 A tent peg is driven into soft ground by a blow from a hammer. The tent peg has mass 0.2 kg and the hammer has mass 3 kg. The hammer strikes the peg vertically.

Immediately before the impact, the speed of the hammer is $16 \, \text{m s}^{-1}$. It is assumed that, immediately after the impact, the hammer and the peg move together vertically downwards.

a Find the common speed of the peg and the hammer immediately after the impact.

Until the peg and hammer come to rest, the resistance exerted by the ground is assumed to be constant and of magnitude R newtons. The hammer and peg are brought to rest 0.05 s after the impact.

b Find, to three significant figures, the value of R. **E**

50 A ball is projected vertically upwards with a speed $u \, \text{m s}^{-1}$ from a point A which is 1.5 m above the ground. The ball moves freely under gravity until it reaches the ground. The greatest height attained by the ball is 25.6 m above A.

a Show that $u = 22.4$.

The ball reaches the ground T seconds after it has been projected from A.

b Find, to two decimal places, the value of T.

The ground is soft and the ball sinks 2.5 cm into the ground before coming to rest. The mass of the ball is 0.6 kg. The ground is assumed to exert a constant resistive force of magnitude F newtons.

c Find, to three significant figures, the value of F.

d State one physical factor which could be taken into account to make the model used in this question more realistic. **E**

51 A particle A, of mass 0.8 kg, resting on a smooth horizontal table, is connected to a particle B, of mass 0.6 kg, which is 1 m from the ground, by a light inextensible string passing over a small pulley at the edge of the table. The particle A is more than 1 m from the edge of the table. The system is released from rest with the horizontal part of the string perpendicular to the edge of the table, the hanging parts vertical and the string taut. Calculate

a the acceleration of A,

b the tension in the string,

c the speed of B when it hits the ground,

d the time taken for B to reach the ground. **E**

52

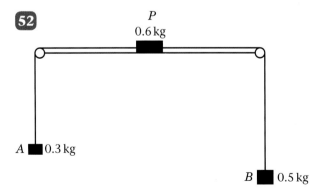

The figure shows a block P of mass 0.6 kg resting on the smooth surface of a horizontal table. Inextensible light strings connect P to blocks A and B which hang freely over light smooth pulleys placed at opposite parallel edges of the table. The masses of A and B are 0.3 kg and 0.5 kg respectively. All portions of the string are taut and perpendicular to their respective edges of the table. The system is released from rest. Calculate

a the common magnitude of the accelerations of the blocks,

b the tensions in the strings. **E**

53 A trailer of mass 600 kg is attached to a car of mass 900 kg by means of a light inextensible tow-bar. The car tows the trailer along a horizontal road. The resistances to motion of the car and trailer are 300 N and 150 N respectively.

a Given that the acceleration of the car and trailer is $0.4\,\mathrm{m\,s^{-2}}$, calculate

  i the tractive force exerted by the engine of the car,

  ii the tension in the tow-bar.

b Given that the magnitude of the force in the tow-bar must not exceed 1650 N, calculate the greatest possible deceleration of the car. **E**

54

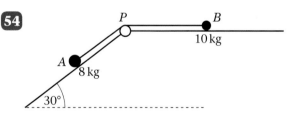

Two particles A and B, of mass 8 kg and 10 kg respectively, are connected by a light inextensible string which passes over a light smooth pulley P. Particle B rests on a smooth horizontal table and A rests on a smooth plane inclined at 30° to the horizontal with the string taut and perpendicular to the line of intersection of the table and the plane as shown in the figure. The system is released from rest. Find

a the magnitude of the acceleration of B,

b the tension in the string,

c the distance covered by B in the first two seconds of motion, given that B does not reach the pulley. **E**

55

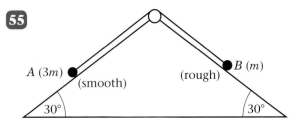

A fixed wedge has two plane faces, each inclined at 30° to the horizontal. Two particles A and B, of mass 3m and m respectively, are attached to the ends of a light inextensible string. Each particle moves on one of the plane faces of the wedge. The string passes over a smooth light pulley fixed at the top of the wedge. The face on which A moves is smooth. The face on which B moves is rough. The coefficient of friction between B and this face is μ. Particle A is held at rest with the string taut. The string lies in the same vertical plane as lines of greatest slope on each plane face of the wedge, as shown in the figure.

The particles are released from rest and start to move. Particle A moves downwards and particle B moves upwards. The acceleration of A and B each have magnitude $\frac{1}{10}g$.

a By considering the motion of A, find, in terms of m and g, the tension in the string.

b By considering the motion of B, find the value of μ.

c Find the resultant force exerted by the string on the pulley, giving its magnitude and direction. **E**

56

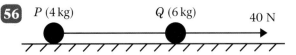

Two particles P and Q, of mass 4 kg and 6 kg respectively, are joined by a light

inextensible string. Initially the particles are at rest on a rough horizontal plane with the string taut. The coefficient of friction between each particle and the plane is $\frac{2}{7}$. A constant force of magnitude 40 N is then applied to Q in the direction PQ, as shown in the figure.

a Show that the acceleration of Q is 1.2 m s^{-2}.

b Calculate the tension in the string when the system is moving.

c State how you have used the information that the string is inextensible.

After the particles have been moving for 7 s, the string breaks. The particle Q remains under the action of the force of magnitude 40 N.

d Show that P continues to move for a further 3 seconds.

e Calculate the speed of Q at the instant when P comes to rest. **E**

57

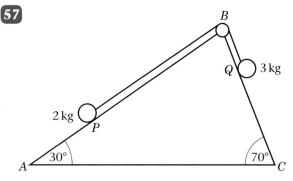

A fixed wedge whose smooth faces are inclined at 30° and 70° to the horizontal has a small smooth pulley fixed on the top edge at B. A light inextensible string, passing over the pulley, has particles P and Q of mass 2 kg and 3 kg respectively attached at its ends. The figure shows a vertical cross-section of the wedge where AB and AC are lines of greatest slope of the faces along which P and Q respectively can slide. The particles are released from rest at time $t = 0$ with the string taut. Assuming that P has not reached B and that Q has not reached C, find

a the distance through which each particle has moved when $t = 0.75\,\text{s}$,

b the tension in the string,

c the magnitude and direction of the resultant force exerted on the pulley by the string.

When $t = 0.75\,\text{s}$ the string breaks and in the subsequent motion P comes to instantaneous rest at time t_1.

d Assuming that P has not reached B, calculate t_1. **E**

58 A car of total mass 1200 kg is moving along a straight horizontal road at a speed of $40\,\text{m s}^{-1}$, when the driver makes an emergency stop. When the brakes are fully applied, they exert a constant force and the car comes to rest after travelling a distance of 80 m. The resistance to motion from all factors other than the brakes is assumed to be constant and of magnitude 500 N.

a Find the magnitude of the force exerted by the brakes when fully applied.

A trailer, with no brakes, is now attached to the car by means of a tow-bar. The mass of the trailer is 600 kg, and when the trailer is moving, it experiences a constant resistance to motion of magnitude 420 N. The tow-bar may be assumed to be a light rigid rod which remains parallel to the road during motion. The car and the trailer come to a straight hill, inclined at an angle α to the horizontal, where $\sin \alpha = \frac{1}{14}$. They move together down the hill. The driver again makes an emergency stop, the brakes applying the same force as when the car was moving along level ground.

b Find the deceleration of the car and the trailer when the brakes are fully applied.

c Find the magnitude of the force exerted on the car by the trailer when the brakes are fully applied.

d Find the maximum speed at which the car and trailer should travel down the hill to ensure that, when the brakes are fully applied, they can stop within 80 m. **E**

59

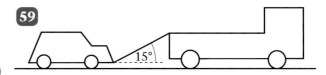

The figure above shows a lorry of mass 1600 kg towing a car of mass 900 kg along a straight horizontal road. The two vehicles are joined by a light tow-bar which is at an angle of 15° to the road. The lorry and the car experience constant resistances to motion of magnitude 600 N and 300 N respectively. The lorry's engine produces a constant horizontal force on the lorry of magnitude 1500 N. Find

a the acceleration of the lorry and the car,

b the tension in the tow-bar.

When the speed of the vehicles is $6\,\text{m s}^{-1}$, the tow-bar breaks. Assuming that the resistance to the motion of the car remains of constant magnitude 300 N,

c find the distance moved by the car from the moment the tow-bar breaks to the moment when the car comes to rest.

d State whether, when the tow-bar breaks, the normal reaction of the road on the car is increased, decreased or remains constant. Give a reason for your answer. **E**

After completing this chapter you should be able to

- find unknown forces acting on a particle which remains at rest in equilibrium

- include relevant forces such as weight, tension and normal reaction in a force diagram.

- consider the magnitude and direction of the friction force and use this together with other forces to solve harder problems involving rough surfaces.

Statics of a particle

You could work out the stability of a man on a roof by modelling the man as a particle and doing calculations based on the angle of slope of the roof and the coefficient of friction between the man and the roof.

Statics is used in engineering for the design of stable structures. It is used in many activities that involve ropes, such as climbing and securing boats. You are applying statics when hanging pictures or wedging open doors.

4.1 You can solve problems about particles in equilibrium by resolving the forces horizontally and vertically.

■ A particle is said to be in equilibrium when it is acted upon by two or more forces and motion does not take place. This means that the resultant of the forces is zero and the particle will remain at rest, or stationary, as it is not subject to acceleration.

■ To solve problems in statics you should:

- Draw a diagram showing clearly the forces acting on the particle.

- Resolve the forces into horizontal and vertical components or, if the particle is on an inclined plane, into components parallel and perpendicular to the plane.

- Set the sum of the components in each direction equal to zero.

- Solve the resulting equations to find the unknown force(s).

> The particle is not accelerating, so $a = 0$.
> $F = ma$
> $\quad = 0$

Example 1

The diagram shows a particle in equilibrium under the forces shown. By resolving horizontally and vertically find the magnitude of the forces P and Q.

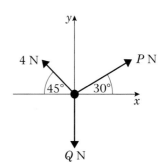

$R(\rightarrow)$

$P \cos 30 - 4 \cos 45 = 0$

$R(\uparrow)$

$P \sin 30 + 4 \sin 45 - Q = 0$

> Resolve horizontally and vertically. Equate the sum of the forces to zero as there is no acceleration (the particle is in equilibrium).

Then solve the equations:

$P = \dfrac{4 \cos 45}{\cos 30} = 3.27$ (3 s.f.)

$Q = P \sin 30 + 4 \sin 45 = 4.46$ (3 s.f.)

> Solve the first equation to find P (as there is only one unknown quantity), and then use your value for P in the second equation to find Q.

> If exact answers are required these would be
>
> $P = \dfrac{4\sqrt{6}}{3}$ and $Q = \dfrac{2(\sqrt{6} + 3\sqrt{2})}{3}$.

Example **2**

The diagram shows a particle in equilibrium under the forces shown.
Find the magnitude of the forces P and Q.

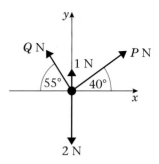

$R(\rightarrow)$

$P \cos 40 - Q \cos 55 = 0$ ①

Resolve horizontally and vertically. Take the direction to the right as positive and to the left as negative.

$R(\uparrow)$

$1 + P \sin 40 + Q \sin 55 - 2 = 0$

$\qquad P \sin 40 + Q \sin 55 = 1$ ②

Take the direction up as positive and down as negative.

Solve these equations as simultaneous equations.

From ①: $P = \dfrac{Q \cos 55}{\cos 40} = 0.749Q$

$0.749 \times \sin 40 = 0.481$
$\qquad \sin 55 = 0.819$

Substituting into ②:

$0.749Q \sin 40 + Q \sin 55 = 1$

$\qquad Q(0.481 + 0.819) = 1$

$1 \div (0.481 + 0.819) = 0.769$

$Q = 0.769$ (3 s.f.)

$P = 0.576$ (3 s.f.)

Substitute the value of Q into equation ① to obtain P.

Example **3**

The diagram shows a particle in equilibrium on an inclined plane under the forces shown.
Find the magnitude of the force P and the magnitude of the angle α.

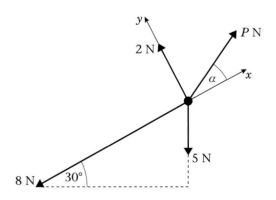

$R(\nearrow)$

$P \cos \alpha - 8 - 5 \sin 30 = 0$ •————

∴ $P \cos \alpha = 8 + 5 \sin 30$ ① •————

Resolve along the plane. Take the direction up the plane as positive.

Rearrange the equation to make $P \cos \alpha$ the subject.

$R(\searrow)$

$P \sin \alpha + 2 - 5 \cos 30 = 0$

∴ $P \sin \alpha = 5 \cos 30 - 2$ ② •————

Resolve perpendicular to the plane. Rearrange the second equation to make $P \sin \alpha$ the subject.

Divide equation ② by equation ① to give:

$\tan \alpha = \dfrac{5 \cos 30 - 2}{8 + 5 \sin 30} = \dfrac{2.330}{10.5} = 0.222$ •————

After division use $\dfrac{\sin \alpha}{\cos \alpha} = \tan \alpha$.

∴ $\alpha = 12.5°$ (3 s.f.) •————

Use $\tan^{-1}$ and give answer to three s.f.

Substitute into equation ①

$P \times \cos 12.5 = 10.5$

∴ $P = 10.8$ (3 s.f.) •————

Substitute your value for α into equation ① to find P.

You could check your answers by substituting into equation ②.

Exercise 4A

Each of the diagrams shows a particle in equilibrium under the action of three or more forces. Using the information given in the diagrams, in each case

a resolve in the x direction,

b resolve in the y direction,

c find the magnitude of any unknown forces (marked P and Q) and the size of any unknown angles (marked θ).

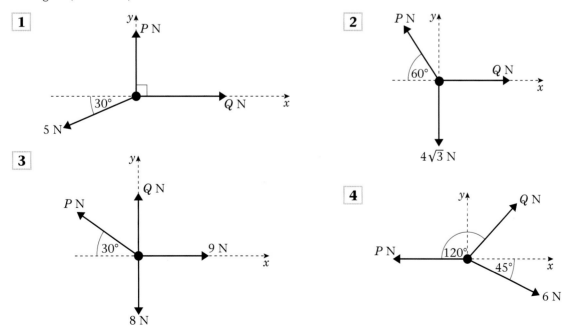

1

2

3

4

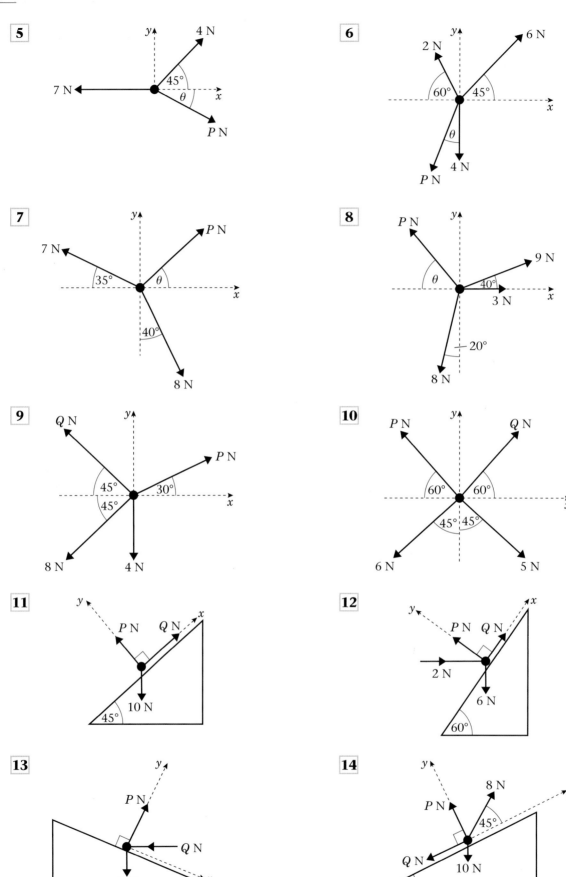

15

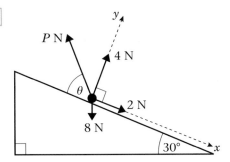

4.2 You need to know when to include additional forces on your diagram, such as weight, tension, thrust, normal reaction and friction.

Example 4

A particle of mass 3 kg is held in equilibrium by two light inextensible strings. One string is horizontal. The other string is inclined at 45° to the horizontal, as shown in the figure. The tension in the horizontal string is $P\,$N and the tension in the other string is $Q\,$N. Find **a** the value of Q, **b** the value of P.

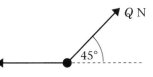

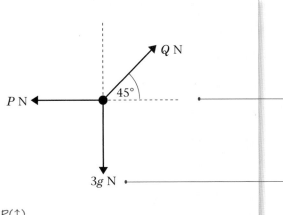

Draw a diagram showing all the forces acting on the particle.

The weight acts vertically down and has magnitude $3g$ where g is taken as $9.8\,\text{m s}^{-2}$.

$R(\uparrow)$

$Q \sin 45 - 3g = 0$

Resolve vertically and put equal to zero.

$\therefore \quad Q = \dfrac{3g}{\sin 45}$

$= 3\sqrt{2}g$

$= 42 \; (2 \text{ s.f.})$

Solve this equation for Q and give answer to two significant figures, as g is only accurate to two significant figures.

$R(\rightarrow)$

$Q \cos 45 - P = 0$

$\therefore \quad P = 3\sqrt{2}g \cos 45°$

Now resolve horizontally and substitute your value for Q.

$= 3g$

$= 29 \; (2 \text{ s.f.})$

Again, answer should only be given to two significant figures.

Example 5

A smooth bead Y is threaded on a light
inextensible string. The ends of the string are
attached to two fixed points X and Z on the same
horizontal level. The bead is held in equilibrium
by a horizontal force of magnitude 8 N acting
parallel to ZX. The bead Y is vertically below X
and $\angle XZY = 30°$ as shown in the figure.

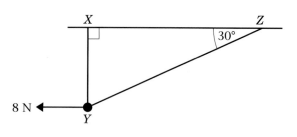

Find **a** the tension in the string, **b** the weight of the bead.

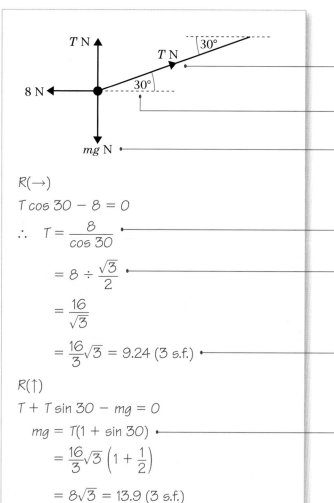

Draw a diagram.

The tension T is the same in both
sections of the string.

This angle is 30° (alternate angles).

Let the weight by mg N.

$R(\rightarrow)$

$T \cos 30 - 8 = 0$

$\therefore \quad T = \dfrac{8}{\cos 30}$

Resolve horizontally and make T the
subject of the formula.

$= 8 \div \dfrac{\sqrt{3}}{2}$

You know that $\cos 30 = \dfrac{\sqrt{3}}{2}$.

$= \dfrac{16}{\sqrt{3}}$

$= \dfrac{16}{3}\sqrt{3} = 9.24$ (3 s.f.)

Give your answer to three significant
figures as an approximation for g has
not been used.

$R(\uparrow)$

$T + T \sin 30 - mg = 0$

$mg = T(1 + \sin 30)$

$= \dfrac{16}{3}\sqrt{3}\left(1 + \dfrac{1}{2}\right)$

Resolve vertically. Make mg the subject
of the formula and substitute T.

$= 8\sqrt{3} = 13.9$ (3 s.f.)

Example 6

A small bag of mass 10 kg is attached at C to the ends of two
light inextensible strings AC and BC. The other ends of the
strings are attached to two fixed points A and B on a horizontal
line. The bag hangs in equilibrium with AC and BC inclined to
the horizontal at 30° and 60° respectively as shown in the figure.
Calculate **a** the tension in AC, **b** the tension in BC.

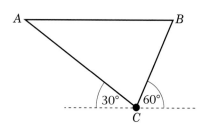

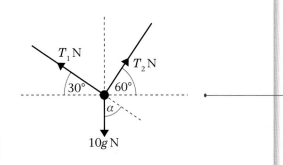

Draw a diagram. Model the bag as a particle and show the forces acting:
- the tension in AC (T_1 N)
- the tension in CB (T_2 N)
- the weight ($10g$ N)

$R(\nwarrow)$

$\qquad T_1 - 10g \cos \alpha = 0$

As $\quad \alpha = 60°,\ T_1 = 5g$

$\qquad\qquad\qquad = 49$ N

Resolve in the direction of T_1.

T_2 has no component in this direction because T_1 and T_2 are perpendicular.

$R(\nearrow)$

$\qquad T_2 - 10g \sin \alpha = 0$

As $\quad \alpha = 60°,\ T_2 = 10g \dfrac{\sqrt{3}}{2}$

$\qquad\qquad\qquad = 5g\sqrt{3}$

$\qquad\qquad\qquad = 85$ N (2 s.f.)

Resolve in the direction of T_2.

Resolving in these directions simplifies the algebra. If you resolve vertically and horizontally you obtain simultaneous equations which you then solve.

Using $g = 9.8$, give answer to two significant figures.

Alternative method

$R(\rightarrow)$

$T_2 \cos 60 - T_1 \cos 30 = 0$

Multiply equation by 2:

$T_2 - T_1\sqrt{3} = 0 \qquad\qquad\qquad ①$

$R(\uparrow)$

$T_2 \sin 60 + T_1 \sin 30 - 10g = 0$

Multiply equation by 2:

$T_2\sqrt{3} + T_1 = 20g \qquad\qquad ②$

$① + ② \times \sqrt{3} \Rightarrow$

$T_2 + 3T_2 - T_1\sqrt{3} + T_1\sqrt{3} = 20g\sqrt{3}$

$\therefore \quad 4T_2 = 20g\sqrt{3}$

$\qquad T_2 = 5g\sqrt{3}.$

Substitute this into equation ①

$5g\sqrt{3} - T_1\sqrt{3} = 0$

$\therefore \qquad\qquad T_1 = 5g$

Resolve horizontally.

You usually resolve horizontally and vertically.

Resolve vertically.

Simplify the simultaneous equations.

Solve to obtain T_1 and T_2.

This method is harder as the algebra is more complicated.

Example **7**

A mass of 3 kg rests on the surface of a smooth plane which is inclined at an angle of 45° to the horizontal. The mass is attached to a cable which passes up the plane along the line of greatest slope and then passes over a smooth pulley at the top of the plane. The cable carries a mass of 1 kg freely suspended at the other end. The masses are modelled as a particle, and the cable as a light inextensible string. There is a force of PN acting horizontally on the 3 kg mass and the system is in equilibrium.

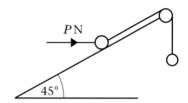

Calculate   **a** the magnitude of P,   **b** the normal reaction between the mass and the plane.

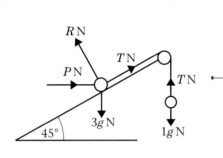

Consider the 1 kg mass:

$R(\uparrow)$

$T - 1g = 0$ ———————————— Resolve vertically to obtain T.

$\therefore \quad T = g = 9.8$

Consider the 3 kg mass: ——————— Resolve up the plane.

$R(\nearrow)$

$T + P\cos 45 - 3g\sin 45 = 0$ ——— R has no component in this direction as R is perpendicular to the plane.

$\therefore \quad P\cos 45 = 3g\sin 45 - T$

But $T = g$ ——————————— Substitute the value for T you found earlier.

$\therefore \quad P\cos 45 = 3g\sin 45 - g$

Divide this equation by $\cos 45$ and use the fact that $\frac{\sin 45}{\cos 45} = \tan 45 = 1$.

$\therefore \quad P = 3g - \dfrac{g}{\cos 45}$

$= 3g - g\sqrt{2}.$ ——————— Use the result that $\cos 45 = \sin 45 = \frac{1}{\sqrt{2}}$.

$= 16 \ (2 \text{ s.f.})$ ——————— Give your answer to two significant figures.

$R(\nwarrow)$ ——————————— Resolve perpendicular to the plane.

$R - P\sin 45 - 3g\cos 45 = 0$

$\therefore \quad R = P\sin 45 + 3g\cos 45$

$= 6g\dfrac{\sqrt{2}}{2} - g$ ——————— Substitute the value of P which you have found to evaluate R.

$= 32 \ (2 \text{ s.f.})$ ——————— Again, give your answer to two significant figures.

Draw a diagram showing the forces acting on each particle. The tension, TN, will be the same throughout the string. The normal reaction, RN acts perpendicular to the plane. Show the weights $3g$N and $1g$N.

Exercise 4B

1 A picture of mass 5 kg is suspended by two light inextensible strings, each inclined at 45° to the horizontal as shown. By modelling the picture as a particle find the tension in the strings when the system is in equilibrium.

2 A particle of mass m kg is suspended by a single light inextensible string. The string is inclined at an angle of 30° to the vertical and the other end of the string is attached to a fixed point O. Equilibrium is maintained by a horizontal force of magnitude 10 N which acts on the particle, as shown in the figure.
Find **a** the tension in the string, **b** the value of m.

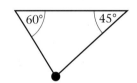

3 A particle of weight 12 N is suspended by a light inextensible string from a fixed point O. A horizontal force of 8 N is applied to the particle and the particle remains in equilibrium with the string at an angle θ to the vertical.
Find **a** the angle θ, **b** the tension in the string.

4 A particle of mass 6 kg hangs in equilibrium, suspended by two light inextensible strings, inclined at 60° and 45° to the horizontal, as shown. Find the tension in each of the strings.

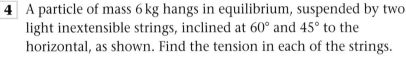

5 A smooth bead B is threaded on a light inextensible string. The ends of the string are attached to two fixed points A and C on the same horizontal level. The bead is held in equilibrium by a horizontal force of magnitude 2 N acting parallel to CA. The sections of string make angles of 60° and 30° with the horizontal.
Find **a** the tension in the string,
    b the mass of the bead.

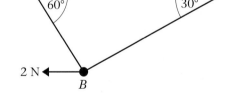

6 A smooth bead B is threaded on a light inextensible string. The ends of the string are attached to two fixed points A and C where A is vertically above C. The bead is held in equilibrium by a horizontal force of magnitude 2 N. The sections AB and BC of the string make angles of 30° and 60° with the vertical respectively.
Find **a** the tension in the string,
    b the mass of the bead, giving your answer to the nearest gramme.

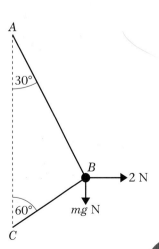

7 A particle of weight 6 N rests on a smooth horizontal surface. It is acted upon by two external forces as shown in the figure. One of these forces is of magnitude 5 N and acts at an angle θ with the horizontal, where $\tan \theta = \frac{4}{3}$. The other has magnitude F N and acts in a horizontal direction. Find

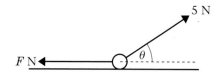

a the value of F,

b the magnitude of the normal reaction between the particle and the surface.

8 A particle of weight 2 N rests on a smooth horizontal surface and remains in equilibrium under the action of the two external forces shown in the figure. One is a horizontal force of magnitude 1 N and the other is a force of magnitude P N at an angle θ to the horizontal, where $\tan \theta = \frac{12}{5}$. Find

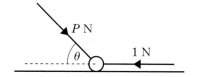

a the value of P,

b the normal reaction between the particle and the surface.

9 A particle A of mass m kg rests on a smooth horizontal table. The particle is attached by a light inextensible string to another particle B of mass $2m$ kg, which hangs over the edge of the table. The string passes over a smooth pulley, which is fixed at the edge of the table so that the string is horizontal between A and the pulley and then is vertical between the pulley and B. A horizontal force F N applied to A maintains equilibrium. The normal reaction between A and the table is R N.

a Find the values of F and R in terms of m.

The pulley is now raised to a position above the edge of the table so that the string is inclined at 30° to the horizontal between A and the pulley. Again the string then hangs vertically between the pulley and B. A horizontal force F' N applied to A maintains equilibrium in this new situation. The normal reaction between A and the table is now R' N.

b Find, in terms of m, the values of F' and R'.

10 A particle of mass 2 kg rests on a smooth inclined plane, which makes an angle of 45° with the horizontal. The particle is maintained in equilibrium by a force P N acting up the line of greatest slope of the inclined plane, as shown in the figure. Find the value of P.

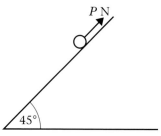

11 A particle of mass 4 kg is held in equilibrium on a smooth plane which is inclined at 45° to the horizontal by a horizontal force of magnitude P N, as shown in the diagram. Find the value of P.

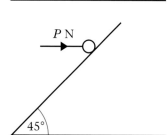

12 A particle A of mass 2 kg rests in equilibrium on a smooth inclined plane. The plane makes an angle θ with the horizontal, where $\tan\theta = \frac{3}{4}$.

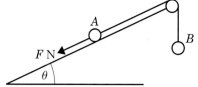

The particle is attached to one end of a light inextensible string which passes over a smooth pulley, as shown in the figure. The other end of the string is attached to a particle B of mass 5 kg. Particle A is also acted upon by a force of magnitude F N down the plane, along a line of greatest slope.

Find **a** the magnitude of the normal reaction between A and the plane, **b** the value of F.

13 A particle Q of mass 5 kg rests in equilibrium on a smooth inclined plane. The plane makes an angle θ with the horizontal, where $\tan\theta = \frac{3}{4}$.

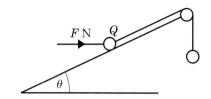

Q is attached to one end of a light inextensible string which passes over a smooth pulley as shown. The other end of the string is attached to a particle of mass 2 kg.

The particle Q is also acted upon by a force of magnitude F N acting horizontally.

Find the magnitude of

a the force F N,   **b** the normal reaction between particle Q and the plane.

14 A particle of weight 20 N rests in equilibrium on a smooth inclined plane. It is maintained in equilibrium by the application of two external forces as shown in the diagram. One of the forces is a horizontal force of 5 N, the other is a force P N acting at 75° to the horizontal.

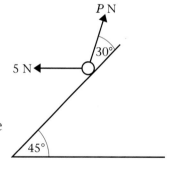

Find   **a** the value of P,

      b the magnitude of the normal reaction between the particle and the plane.

4.3 You can solve statics problems involving friction by using the relationship $F \leqslant \mu R$.

The force of **friction** was introduced in Chapter 3. In that chapter you solved dynamics problems where the frictional force acted in a direction opposing motion and took its limiting value μR, where μ was the coefficient of friction and R was the normal reaction force.

In statics you solve problems where there is no motion.
When a body is in equilibrium under the action of a number of forces including friction you need to consider whether the body is on the point of moving or not.

■ **The maximum value of the frictional force $F_{\text{MAX}} = \mu R$ is reached when the body you are considering is on the point of moving. The body is then said to be in limiting equilibrium.**

In many cases the force of friction will be less than μR, as a smaller force is sufficient to prevent motion and to maintain equilibrium. In these situations the equilibrium is not limiting.

■ In general the force of friction F is such that $F \leqslant \mu R$, and the direction of the friction force is opposite to the direction in which the body would move if the friction force were absent.

Example 8

A block of mass 3 kg rests on a rough horizontal plane. The coefficient of friction between the block and the plane is 0.4. When a horizontal force P N is applied to the block the block remains in equilibrium.

a Find the value of P for which the equilibrium is limiting.

b Find the value of F when $P = 8$ N.

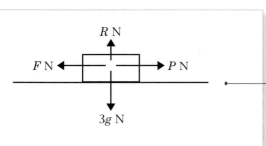

Draw a diagram showing the weight, normal reaction, friction and the force P N. Model the block as a particle.

$R(\uparrow)$ — Resolve vertically to find R.

$R - 3g = 0$

$\therefore \quad R = 3g$

a As equilibrium is limiting:

$F = \mu R$ — This gives the limiting friction.

$\therefore \quad F = 0.4 \times 3g = 11.76$

$R(\rightarrow)$

$P - F = 0$ — Resolving horizontally gives the value of P.

$\therefore \quad P = 12$ (2 s.f.)

b When $P = 8$ — As 8 N is less than 11.76 N equilibrium is maintained. Friction is not limiting and $F < \mu R$.

$R(\rightarrow)$

$P - F = 0$

$\therefore \quad F = 8$

Example 9

A mass of 8 kg rests on a rough horizontal plane. The mass may be modelled as a particle, and the coefficient of friction between the mass and the plane is 0.5. Find the magnitude of the maximum force P N which acts on this mass without causing it to move if

a the force P is horizontal,

b the force P acts at an angle $60°$ above the horizontal.

a

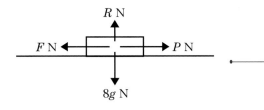

Draw a diagram showing the weight $8g$ N, the normal reaction R N, the force P N and the friction F N. The friction is in the opposite direction to force P N.

This is an example of limiting equilibrium

$R(\uparrow)$

$R - 8g = 0$

$\therefore \quad R = 8g$

As the question asks you for the maximum force before movement takes place.

As friction is limiting, $F = \mu R$

$\therefore \quad F = 0.5 \times 8g$

$\qquad = 39.2$

This is the condition for limiting equilibrium.

$R(\rightarrow)$

$\quad P - F = 0$

$\therefore \qquad P = F = 39 \ (2 \text{ s.f.})$

Give your answer to two significant figures.

b

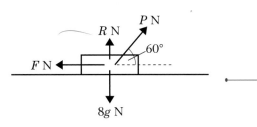

Draw another diagram showing P at $60°$ above the horizontal.

Again this is limiting equilibrium.

$R(\uparrow)$

$R + P \sin 60 - 8g = 0$

$\therefore \quad R = 8g - P \sin 60$

Express R in terms of P.

As friction is limiting, $F = \mu R$

$\therefore \quad F = 0.5 (8g - P \sin 60)$

Use $F = \mu R$ with $\mu = 0.5$.

$R(\rightarrow)$

$P \cos 60 - F = 0$

$\therefore P \cos 60 = 0.5 (8g - P \sin 60)$

As $F = P \cos 60$ eliminate F from the previous equation.

$\therefore \quad P \cos 60 + 0.5 P \sin 60 = 0.5 \times 8g$

$\therefore \quad P(\cos 60 + 0.5 \sin 60) = 4g$

Collect the terms in P and factorise to make P the subject.

$\therefore \quad P = \dfrac{4g}{\cos 60 + 0.5 \sin 60}$

$\qquad P = 42 \ (2 \text{ s.f.})$

Example **10**

A box of mass 10 kg rests in limiting equilibrium on a rough plane inclined at 20° above the horizontal. Find the coefficient of friction between the box and the plane.

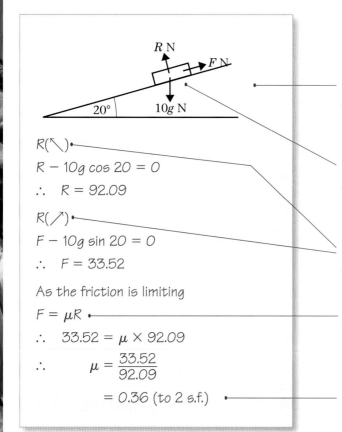

Model the box as a particle and draw a diagram showing the weight, the normal reaction and the force of friction.

$R(\nwarrow)$

$R - 10g \cos 20 = 0$

$\therefore \quad R = 92.09$

The friction acts up the plane, as it acts in an opposite direction to the motion that would take place if there was no friction.

$R(\nearrow)$

$F - 10g \sin 20 = 0$

$\therefore \quad F = 33.52$

Resolve perpendicular and parallel to the plane.

As the friction is limiting

$F = \mu R$

Find R and F, then use $F = \mu R$ to find μ.

$\therefore \quad 33.52 = \mu \times 92.09$

$\therefore \qquad \mu = \dfrac{33.52}{92.09}$

$\qquad = 0.36 \text{ (to 2 s.f.)}$

Give your answer to two significant figures and note that $\mu = \tan 20°$.

Example **11**

A parcel of mass 2 kg is placed on a rough plane inclined at an angle α to the horizontal where $\sin \alpha = \frac{5}{13}$. Given that the parcel may be modelled as a particle and that the coefficient of friction is $\frac{1}{3}$, find the magnitude of the force P N acting along the plane which is just sufficient to prevent the particle from

a sliding up the plane,

b sliding down the plane.

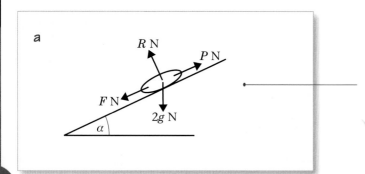

Draw a diagram showing the forces acting on the particle.
The force of friction acts down the plane as it prevents motion up the plane.

$R(\nwarrow)$

$R - 2g \cos \alpha = 0$

$\therefore \quad R = 2g \times \dfrac{12}{13}$

$R(\nearrow)$

$P - F - 2g \sin \alpha = 0$

$\therefore \quad F = P - 2g \times \dfrac{5}{13}$

As the friction is limiting, $F = \mu R$:

$\therefore \quad P - 2g \times \dfrac{5}{13} = \dfrac{1}{3} \times 2g \times \dfrac{12}{13}$

$\therefore \quad P = \dfrac{8g}{13} + \dfrac{10g}{13}$

$\therefore \quad P = \dfrac{18g}{13} = 14$ (to 2 s.f.)

Resolve perpendicular to and parallel to the plane, using

$\sin \alpha = \dfrac{5}{13}$ and

$\cos \alpha = \dfrac{12}{13}$.

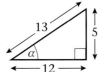

Find R and find F in terms of P, then use $F = \mu R$, with $\mu = \dfrac{1}{3}$.

Find P and give your answer to two significant figures as the approximation for g was used in the calculation.

b

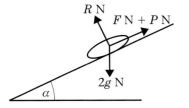

In this example F acts up the plane.

$R(\nwarrow)$

$R = 2g \times \dfrac{12}{13}$ as in part **a**

$R(\nearrow)$

$P + F - 2g \sin \alpha = 0$

$\therefore \quad F = 2g \times \dfrac{5}{13} - P$

Again the friction is limiting so $F = \mu R$:

$\therefore \quad 2g \times \dfrac{5}{13} - P = \dfrac{1}{3} \times 2g \times \dfrac{12}{13}$

$\therefore \quad P = \dfrac{10g}{13} - \dfrac{8g}{13}$

$\quad = \dfrac{2g}{13} = 1.5 \text{ N}$ (to 2 s.f.)

Resolve perpendicular to and parallel to the plane.

Note that '$-F$' in part **a** becomes '$+F$' in part **b**.

Again use $F = \mu R$ with $\mu = \dfrac{1}{3}$.

Give your answer to two significant figures.

Example **12**

A box of mass 1.6 kg is placed on a rough plane, which is inclined at 45° to the horizontal. The box is kept in equilibrium by a light string, which makes an angle of 15° with the line of greatest slope of the inclined plane as shown in the diagram.

When the tension in the string is 15 N, the box is in limiting equilibrium and about to move up the plane.

a Find the value of the coefficient of friction between the box and the plane.

The tension in the string is then reduced to 10 N.

b Determine the magnitude and direction of the force of friction in this case.

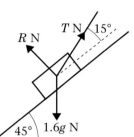

a

$R(\nwarrow)$

$R + T\sin 15 - 1.6g\cos 45 = 0$

$\therefore\ R = 1.6g\cos 45 - T\sin 15$

Draw a diagram showing the forces acting, with friction acting down the plane.

$R(\nearrow)$

$T\cos 15 - F - 1.6g\sin 45 = 0$

$\therefore\ F = T\cos 15 - 1.6g\sin 45$

Resolve perpendicular to the plane, then make R the subject.

When $T = 15$, friction is limiting, so $F = \mu R$:

$\therefore\ 15\cos 15 - 1.6g\sin 45 = \mu(1.6g\cos 45 - 15\sin 15)$

$\therefore\ \mu = \dfrac{15\cos 15 - 1.6g\sin 45}{1.6g\cos 45 - 15\sin 15}$

$= \dfrac{3.40145\ldots}{7.2051\ldots}$

$\therefore\ \mu = 0.47$ (to 2 s.f.)

Resolve parallel to the plane, then make F the subject.

Substitute $T = 15$ and use $F = \mu R$.

Use the previous expression for R and substitute $T = 10$.

b When $T = 10$

$R = 1.6g\cos 45 - 10\sin 15$

$= 8.499$

$\therefore\ \mu R = 4.012$

and $F = 10\cos 15 - 1.6g\sin 45$

$= -1.428$

As $|F| < \mu R$ the friction is not limiting and as $F < 0$ the friction force acts up the plane, with magnitude 1.4 N (2 s.f.)

Calculate μR with the value for μ found in part **a**.

Use your expression for F and substitute $T = 10$.

The negative value implies that F acts up the plane in this situation.

As $1.428 < 4.012$ the friction is not limiting.

Exercise 4C

1 A book of mass 2 kg rests on a rough horizontal table. When a force of magnitude 8 N acts on the book, at an angle of 20° to the horizontal in an upward direction, the book is on the point of slipping.

Calculate, to three significant figures, the value of the coefficient of friction between the book and the table.

2 A block of mass 4 kg rests on a rough horizontal table. When a force of 6 N acts on the block, at an angle of 30° to the horizontal in a downward direction, the block is on the point of slipping. Find the value of the coefficient of friction between the block and the table.

3 A block of weight 10 N is at rest on a rough horizontal surface. A force of magnitude 3 N is applied to the block at an angle of 60° above the horizontal in an upward direction. The coefficient of friction between the block and the surface is 0.3.

a Calculate the force of friction, **b** determine whether the friction is limiting.

4 A packing crate of mass 10 kg rests on rough ground. It is filled with books which are evenly distributed through the crate. The coefficient of friction between the crate and the ground is 0.3.

a Find the mass of the books if the crate is in limiting equilibrium under the effect of a horizontal force of magnitude 147 N.

b State what modelling assumptions you have made.

5 A block of mass 2 kg rests on a rough horizontal plane. A force P acts on the block at an angle of 45° to the horizontal. The equilibrium is limiting, with $\mu = 0.3$.

Find the magnitude of P if

a P acts in a downward direction, **b** P acts in an upward direction.

6 A particle P of mass 0.3 kg is on a rough plane which is inclined at an angle 30° to the horizontal. The particle is held at rest on the plane by a force of magnitude 3 N acting up the plane, in a direction parallel to a line of greatest slope of the plane. The particle is on the point of slipping up the plane. Find the coefficient of friction between P and the plane.

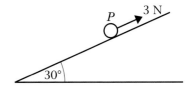

7 A particle of mass 1.5 kg rests in equilibrium on a rough plane under the action of a force of magnitude X N acting up a line of greatest slope of the plane. The plane is inclined at 25° to the horizontal. The particle is in limiting equilibrium and on the point of moving up the plane. The coefficient of friction between the particle and the plane is 0.25.

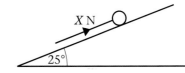

Calculate **a** the normal reaction of the plane on P, **b** the value of X.

8 A horizontal force of magnitude 20 N acts on a block of mass 1.5 kg, which is in equilibrium resting on a rough plane inclined at 30° to the horizontal. The line of action of the force is in the same vertical plane as the line of greatest slope of the inclined plane.

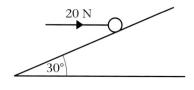

a Find the magnitude and direction of the frictional force acting on the block.

b Find the normal reaction between the block and the plane.

c What can you deduce about the coefficient of friction between the block and the plane?

9 A box of mass 3 kg lies on a rough plane inclined at 40° to the horizontal. The box is held in equilibrium by means of a horizontal force of magnitude X N. The line of action of the force is in the same vertical plane as the line of greatest slope of the inclined plane. The coefficient of friction between the box and the plane is 0.3 and the box is in limiting equilibrium and is about to move up the plane.

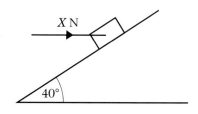

a Find the normal reaction between the box and the plane.   **b** Find X.

10 A small child, sitting on a sledge, rests in equilibrium on an inclined slope. The sledge is held by a rope which lies along the slope and is under tension. The sledge is on the point of slipping down the plane. Modelling the child and sledge as a particle and the rope as a light inextensible string, calculate the tension in the rope, given that the mass of the child and sledge is 22 kg, the coefficient of friction is 0.125 and that the slope is a plane inclined at 35° with the horizontal. The direction of the rope is along a line of greatest slope of the plane.

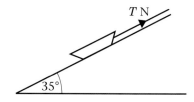

11 A box of mass 0.5 kg is placed on a plane which is inclined at an angle of 40° to the horizontal. The coefficient of friction between the box and the plane is $\frac{1}{5}$. The box is kept in equilibrium by a light string which lies in a vertical plane containing a line of greatest slope of the plane. The string makes an angle of 20° with the plane, as shown in the diagram. The box is in limiting equilibrium and may be modelled as a particle. The tension in the string is T N. Find T

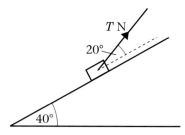

a if the box is about to move up the plane,

b if the box is about to move down the plane.

12 A box of mass 1 kg is placed on a plane, which is inclined at an angle of 40° to the horizontal. The box is kept in equilibrium by a light string, which lies in a vertical plane containing a line of greatest slope of the plane. The string makes an angle of 20° with the plane, as shown in the diagram. The box is in limiting equilibrium and may be modelled as a particle. The tension in the string is 10 N and the coefficient of friction between the box and the plane is μ. Find μ if the box is about to move up the plane.

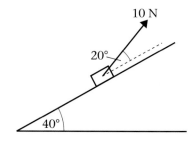

Mixed exercise 4D

1 A particle *P* of mass 2 kg is held in equilibrium under gravity by two light inextensible strings. One string is horizontal and the other is inclined at an angle *α* to the horizontal, as shown in the diagram. The tension in the horizontal string is 15 N. The tension in the other string is *T* newtons.

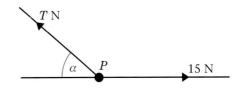

a Find the size of the angle *α*.　　**b** Find the value of *T*.　　**E**

2 A particle is suspended by two light inextensible strings and hangs in equilibrium. One string is inclined at 30° to the horizontal and the tension in that string is of magnitude 40 N. The second string is inclined at 60° to the horizontal. Calculate in N

a the weight of the particle,　　**b** the magnitude of the tension in the second string.　　**E**

3 A particle *P* of weight 6 N is attached to one end of a light inextensible string. The other end of the string is attached to a fixed point *O*. A horizontal force of magnitude *F* newtons is applied to *P*. The particle *P* is in equilibrium under gravity with the string making an angle of 30° with the vertical, as shown in the diagram.

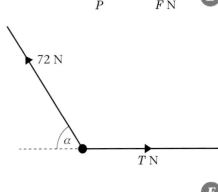

Find, to three significant figures,

a the tension in the string,　　**b** the value of *F*.　　**E**

4 A body of mass 5 kg is held in equilibrium under gravity by two inextensible light ropes. One rope is horizontal, the other is at an angle *α* to the horizontal, as shown in the diagram.

The tension in the rope inclined at *α* to the horizontal is 72 N. Find

a the angle *α*, giving your answer to the nearest degree,

b the value of *T*, to the nearest whole number.　　**E**

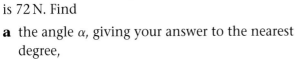

5 In the diagram, ∠*AOC* = 90° and ∠*BOC* = *θ*°. A particle at *O* is in equilibrium under the action of three coplanar forces. The three forces have magnitudes 8 N, 12 N and *X* N and act along *OA*, *OB* and *OC* respectively. Calculate

a the value, to one decimal place, of *θ*,

b the value, to two decimal places, of *X*.

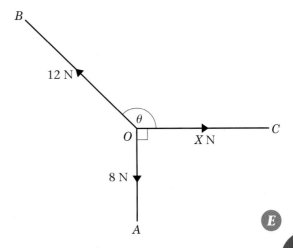

6 The two ends of a string are attached to two points A and B of a horizontal beam. A package of mass 2 kg is attached to the string at the point C. When the package hangs in equilibrium $\angle BAC = 20°$ and $\angle ABC = 40°$, as shown below.

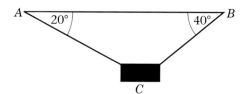

By modelling the package as a particle and the string as light and inextensible, find, to three significant figures,

a the tension in AC, **b** the tension in BC. **E**

7 A block of mass 3 kg rests on a rough, horizontal table. When a force of magnitude 10 N acts on the block at an angle of 60° to the horizontal in an upwards direction, the block is on the point of slipping. Calculate, to two significant figures, the value of the coefficient of friction between the block and the table. **E**

8 A particle P of mass $7m$ is placed on a rough horizontal table, the coefficient of friction between P and the table being μ. A force of magnitude $2mg$, acting upwards at an acute angle α to the horizontal, is applied to P and equilibrium is on the point of being broken by the particle sliding on the table. Given that $\tan \alpha = \frac{5}{12}$, find the value of μ. **E**

9 A box of mass 50 kg rests on rough horizontal ground. The coefficient of friction between the box and the ground is 0.6. A force of magnitude P newtons is applied to the box at an angle of 15° to the horizontal, as shown in the diagram, and the box is now in limiting equilibrium. By modelling the box as a particle find, to three significant figures, the value of P. **E**

10 A book of mass 2 kg rests on a rough plane inclined at an angle $\alpha°$ to the horizontal. Given that the coefficient of friction between the book and the plane is 0.2, and that the book is on the point of slipping down the plane, find, to the nearest degree, the value of α. **E**

11 a A book is placed on a desk lid which is slowly tilted. Given that the book begins to slide when the inclination of the lid to the horizontal is 30°, find the coefficient of friction between the book and the desk lid.

b State an assumption you have made about the book when forming the mathematical model you used to solve part **a**. **E**

12 A particle is placed on a smooth plane inclined at 35° to the horizontal. The particle is kept in equilibrium by a horizontal force, of magnitude 8 N, acting in the vertical plane containing the line of greatest slope of the inclined plane through the particle. Calculate, in N to one decimal place,

a the weight of the particle,

b the magnitude of the force exerted by the plane on the particle. **E**

13 A particle of mass 0.3 kg lies on a smooth plane inclined at an angle α to the horizontal, where $\tan \alpha = \frac{3}{4}$. The particle is held in equilibrium by a horizontal force of magnitude Q newtons. The line of action of this force is in the same vertical plane as a line of greatest slope of the inclined plane. Calculate the value of Q, to one decimal place. **E**

14 A small boat B of mass 100 kg is standing on a ramp which is inclined at 18° to the horizontal. A force of magnitude 400 N is applied to B and acts down the ramp as shown in the diagram. The boat is in limiting equilibrium on the point of sliding down the ramp. Find the coefficient of friction between B and the ramp, giving your answer to two decimal places. **E**

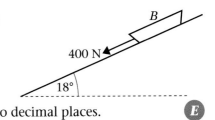

15 A parcel of mass 5 kg lies on a rough plane inclined at an angle of α to the horizontal, where $\tan \alpha = \frac{3}{4}$. The parcel is held in equilibrium by the action of a horizontal force of magnitude 20 N, as shown in the diagram. The force acts in a vertical plane through a line of greatest slope of the plane.
The parcel is on the point of sliding down the plane.
Find the coefficient of friction between the parcel and the plane. **E**

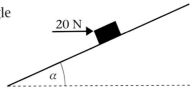

16 A small parcel of mass 3 kg is held in equilibrium on a rough plane by the action of a horizontal force of magnitude 30 N acting in a vertical plane through a line of greatest slope. The plane is inclined at an angle of 30° to the horizontal, as shown in the diagram.

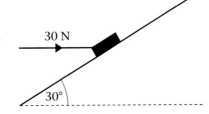

The parcel is modelled as a particle. The parcel is on the point of moving up the slope.

a Draw a diagram showing all the forces acting on the parcel.

b Find the normal reaction on the parcel.

c Find the coefficient of friction between the parcel and the plane. **E**

17 A box of mass 6 kg lies on a rough plane inclined at an angle of 30° to the horizontal. The box is held in equilibrium by means of a horizontal force of magnitude P newtons, as shown in the diagram.

The line of action of the force is in the same vertical plane as a line of greatest slope of the plane. The coefficient of friction between the box and the plane is 0.4.

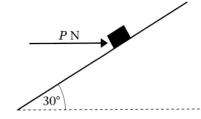

The box is modelled as a particle. Given that the box is in limiting equilibrium and on the point of moving up the plane, find,

a the normal reaction exerted on the box by the plane,

b the value of P.

The horizontal force is removed.

c Show that the box will now start to move down the plane. **E**

18 A box of mass 1.5 kg is placed on a plane which is inclined at an angle of 30° to the horizontal. The coefficient of friction between the box and plane is $\frac{1}{3}$. The box is kept in equilibrium by a light string which lies in a vertical plane containing a line of greatest slope of the plane. The string makes an angle of 20° with the plane, as shown in the diagram.

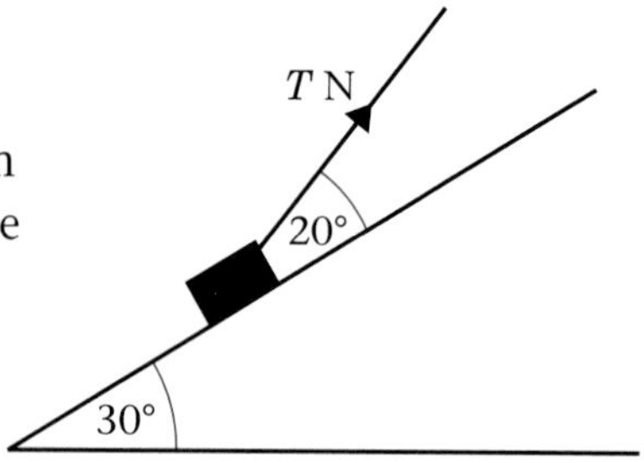

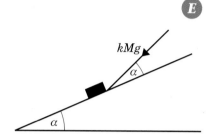

The box is in limiting equilibrium and is about to move up the plane. The tension in the string is T newtons. The box is modelled as a particle. Find the value of T.

E

19 A rough slope is inclined at an angle α to the horizontal, where $\alpha < 45°$. A small parcel of mass M is at rest on the slope, and the coefficient of friction between the parcel and the slope is μ. A force of magnitude kMg, where k is a constant, is applied to the parcel in a direction making an angle α with a line of greatest slope, as shown in the diagram.

The line of action of the force is in the same vertical plane as the line of greatest slope.

Given that the parcel is on the point of moving down the slope, show that:

$$k = \frac{\mu \cos \alpha - \sin \alpha}{\cos \alpha - \mu \sin \alpha}$$

E

20 A parcel A of mass 2 kg rests on a rough slope inclined at an angle θ to the horizontal, where $\tan \theta = \frac{3}{4}$. A string is attached to A and passes over a small smooth pulley fixed at P. The other end of the string is attached to a weight B of mass 2.2 kg, which hangs freely, as shown in the diagram.

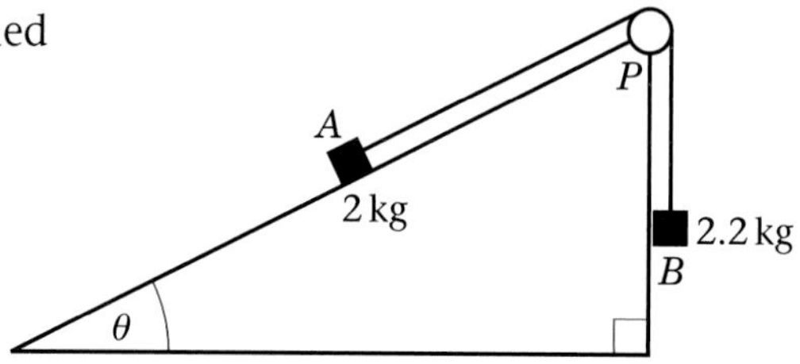

The parcel A is in limiting equilibrium and about to slide up the slope. By modelling A and B as particles and the string as light and inextensible, find

a the normal contact force acting on A,

b the coefficient of friction between A and the slope.

E

21 A light inextensible string passes over a smooth peg, and is attached at one end to a particle of mass m kg and at the other end to a ring also of mass m kg. The ring is threaded on a rough vertical wire as shown in the diagram. The system is in limiting equilibrium with the part of the string between the ring and the peg making an angle of 60° with the vertical wire.

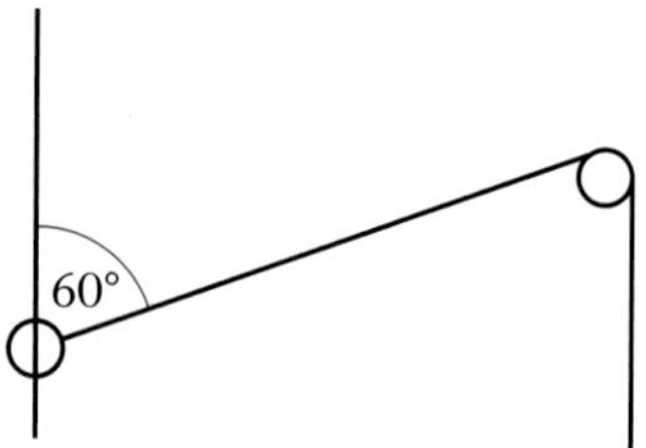

Calculate the coefficient of friction between the ring and the wire, giving your answer to three significant figures.

22 A light inextensible string passes over a smooth peg, and is attached at one end to a particle of mass $3m$ kg and at the other end to a ring of mass $2m$ kg. The ring is threaded on a rough vertical wire as shown in the diagram. The system is in limiting equilibrium with the part of the string between the ring and the peg making an angle of 30° with the vertical wire.

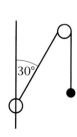

Calculate the coefficient of friction between the ring and the wire, giving your answer to three significant figures.

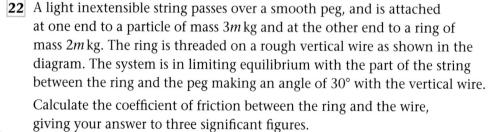

23 A ring of mass 0.3 kg is threaded on a fixed, rough horizontal curtain pole. A light inextensible string is attached to the ring. The string and the pole lie in the same vertical plane. The ring is pulled downwards by the string which makes an angle α to the horizontal, where $\tan \alpha = \frac{1}{4}$ as shown in the diagram.

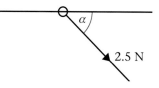

The tension in the string is 2.5 N.

Given that, in this position, the ring is in limiting equilibrium,

a find the coefficient of friction between the ring and the pole.

The direction of the string is now altered so that the ring is pulled upwards. The string lies in the same vertical plane as before and again makes an angle α with the horizontal, as shown in the diagram below.

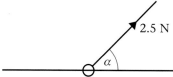

The tension in the string is again 2.5 N.

b Find the normal reaction exerted by the pole on the ring.

c State whether the ring is in equilibrium in the position shown in the second figure, giving a brief justification for your answer. You need make no further detailed calculation of the forces acting.

E

Summary of key points

1 A particle is said to be in equilibrium when it is acted upon by two or more forces and motion does not take place. This means that the resultant of the forces is zero and the particle will remain at rest or stationary, as it is not subject to acceleration.

2 To solve problems in statics you should:
- draw a diagram showing clearly the forces acting on the particle(s)
- resolve the forces into horizontal and vertical components or, if the particle is on an inclined plane, into components parallel and perpendicular to the plane
- set the sum of the components in each direction equal to zero
- solve the resulting equations to find the unknown force(s) and angle(s).

3 You can add forces of weight, tension, thrust, normal reaction and friction to a force diagram as appropriate.

4 The maximum value of the frictional force $F_{\text{MAX}} = \mu R$ is reached when the body you are considering is on the point of moving. The body is then said to be in limiting equilibrium.

5 In general the force of friction F is such that $F \leq \mu R$, where R is the normal reaction.

6 The direction of the friction force is opposite to the direction in which the body would move if the friction force were absent.

5

After completing this chapter you should be able to

- calculate the magnitude of the turning effect of a force applied to a rigid body
- solve problems about bodies in equilibrium
- solve problems about non-uniform bodies.

Moments

One person sitting at the end of a see-saw can balance the weight of two people sitting closer in. The turning effect produced by a force depends on its distance from the pivot point.

It is easier to close the door if you push at the edge. This is because you are applying force further away from the pivot point (the hinge). You would need a greater force closer to the hinge to produce the same turning effect.

5.1 You can find the moment of a force acting on a body.

■ **The moment of a force measures the turning effect of the force on the body on which it is acting.**

The moment of a force depends on the magnitude of the force and its distance from the axis of rotation.

■ **The moment of a force F about a point P is the product of the magnitude of the force and the perpendicular distance of the line of action of the force from the point P.**

Example 1

Find the moment of F about P.

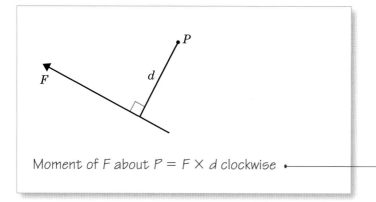

Moment of F about $P = F \times d$ clockwise • ——————— When describing a moment you need to give a direction of rotation.

Example 2

Find the moment of F about P.

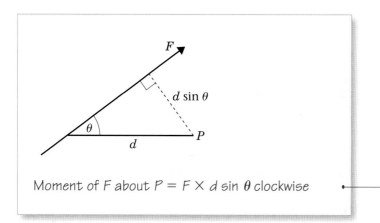

Moment of F about $P = F \times d \sin \theta$ clockwise • ——————— Use trigonometry to find the perpendicular distance.

■ **The magnitude of the force is measured in newtons (N) and the distance is measured in metres (m), so the moment of the force is measured in newton-metres (N m).**

Example 3

The diagram shows two forces acting on a lamina. Find the moment of each of the forces about the point P.

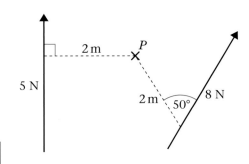

The distance given on the diagram is the perpendicular distance, so you can substitute the given values directly into the formula.

Don't forget to include the sense of the rotation when you describe the moment of the force.

Moment of the 5 N force

= magnitude of force × perpendicular distance

= 5 × 2 = 10 N m clockwise

Moment of the 8 N force

= magnitude of force × perpendicular distance

= 8 × 2 sin 50° = 12.3 N m anticlockwise (3 s.f.)

This time you need to start by finding the perpendicular distance 2 sin 50°.

The turning effect is in the opposite direction.

Exercise 5A

Calculate the moment about P of each of these forces acting on a lamina.

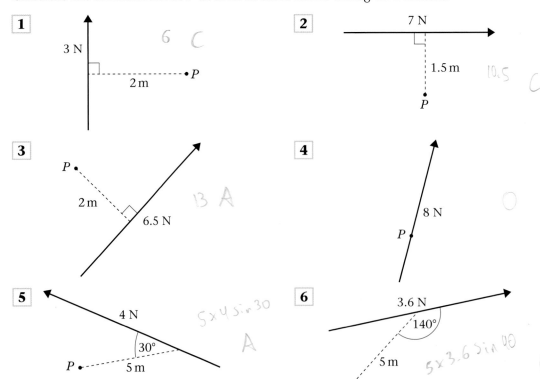

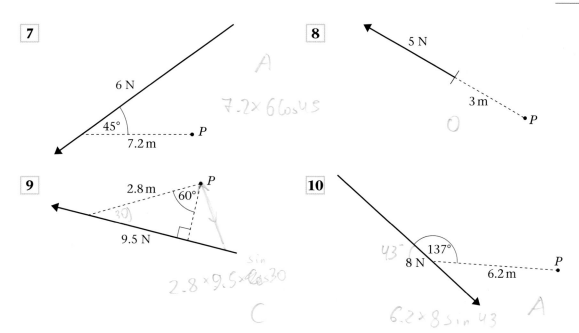

7

6 N

45°

7.2 m

•P

A

7.2 × 6 cos 45

8

5 N

3 m

•P

O

9

2.8 m

60°

P

9.5 N

30

2.8 × 9.5 × cos 30

sin

C

10

137°

43°

8 N

6.2 m

P

6.2 × 8 sin 43

A

5.2 You can find the sum of the moments of a set of forces acting on a body.

■ When you have several coplanar forces acting on a body, you can add the moments about a point. You need to choose a positive direction (clockwise or anticlockwise) and consider the sense of rotation of each moment.

Example 4

The diagram shows a set of forces acting on a light rod. Calculate the sum of the moments about the point *P*.

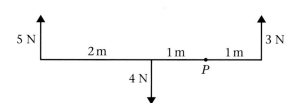

5 N

2 m

1 m

1 m

3 N

4 N

P

The moment of the 5 N force is

5 × (2 + 1) = 15 N m clockwise.

The moment of the 4 N force is

4 × 1 = 4 N m anticlockwise.

The moment of the 3 N force is

3 × 1 = 3 N m anticlockwise.

Total clockwise = 15 N m

Total anticlockwise = 4 + 3 = 7 N m

∴ The sum of the moments is

15 − 7 = 8 N m clockwise

Find the moment of each force.

Add together the moments acting in the same direction.

The clockwise total is greater than the anticlockwise total, so choose clockwise as the positive direction.

Example 5

The diagram shows two forces acting on a lamina.
Calculate the sum of the moments of the forces
about the point P.

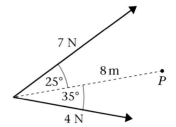

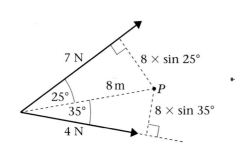

The moment of the 7 N force

$= 7 \times 8 \sin 25° = 23.66... \text{ N m clockwise.}$

The moment of the 4 N force

$= 4 \times 8 \sin 35° = 18.35... \text{ N m anticlockwise.}$

$\therefore$ Sum of the moments

$= 23.66... - 18.35...$

$= 5.31 \text{ N m clockwise}$

Draw a diagram and find the perpendicular distances.

Find the moments of both forces.

The clockwise moment is bigger so choose this as the positive direction.

Add the unrounded values then round your answer to three significant figures.

■ **When the distance given is not the perpendicular distance of the line of action from the pivot you can find the moment by resolving the force into components.**

Example 6

The diagram shows a force acting on a lamina.
Find the moment of the force about the point P.

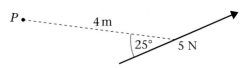

Method 1

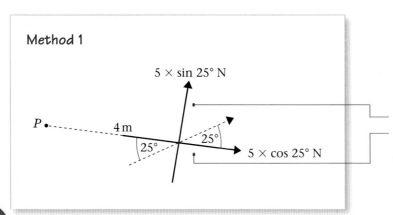

You can resolve the 5 N force into two components: one acting perpendicular to the 4 m distance you were given and one acting parallel to it.

Moment of the perpendicular component of force
($5 \times \sin 25°$ N) about P

$= 4 \times 5 \sin 25° = 20 \times \sin 25°$ •————— Using $F \times d$.

$= 8.45$ N m anticlockwise (3 s.f.)

The moment of the parallel component of force
($5 \times \cos 25°$ N) about P is zero, since the line of
action of this force passes through P and the
perpendicular distance is therefore zero.

> This is an important result which
> you can use to help you solve
> more complicated problems
> later on.

Therefore the moment of the 5 N force about P is
8.45 N m (3 s.f.) anticlockwise.

Method 2

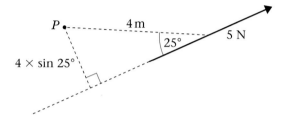

Moment of the force

$= 5 \times 4 \sin 25°$

$= 20 \sin 25°$

$= 8.45$ N m anticlockwise (3 s.f.)

Exercise 5B

1 These diagrams show sets of forces acting on a light rod. For each rod, calculate the sum of
the moments about P.

a
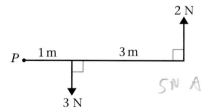
$5N$ A

b
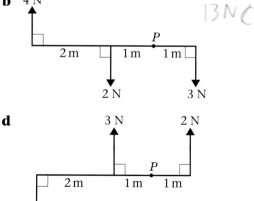
$13NC$

c
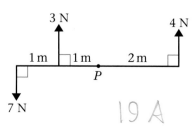
19 A

d
9 A

e

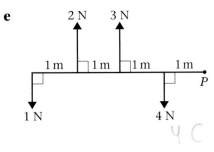

f

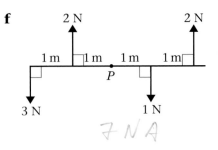

2 These diagrams show forces acting on a lamina. In each case, find the sum of the moments of the set of forces about *P*.

a

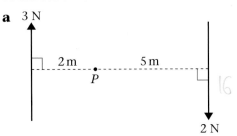

b

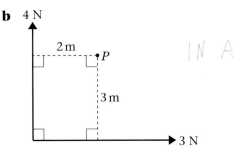

c

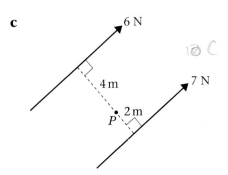

d

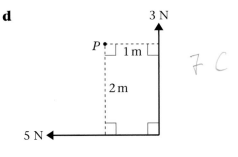

e

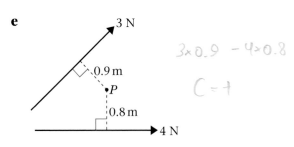

f

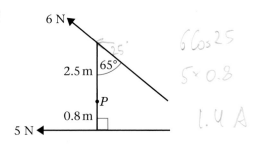

5.3 You can solve problems about bodies resting in equilibrium by equating the clockwise and anticlockwise moments.

When a particle is in equilibrium under the action of a set of forces the resultant force in any direction is zero. The same is true for a rigid body. In addition, if a rigid body is in equilibrium under the action of a set of forces then the body will not turn about any point.

■ **When a rigid body is in equilibrium the resultant force in any direction is zero, and the sum of the moments about any point is zero.**

Example 7

The diagram shows a uniform rod AB of length
3 m and weight 20 N resting horizontally on
supports at A and C, where $AC = 2$ m.
Calculate the magnitude of the reaction at each of the supports.

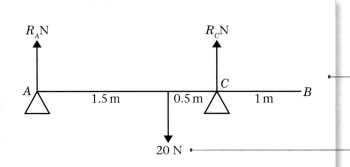

The rod is in equilibrium.

Resolving vertically:

$R_A + R_C = 20$

Considering the moments about point A:

$20 \times 1.5 = R_C \times (1.5 + 0.5)$,

$30 = 2R_C$

$15 = R_C$

$R_A + 15 = 20$

$R_A = 5$

Therefore the reaction at A is 5 N and the
reaction at C is 15 N.

Draw a diagram showing all the forces acting.

The weight of the rod acts at its centre of mass. You are told that this is a uniform rod, so the weight acts at the mid-point of the rod.

Total of forces acting upwards = total of forces acting downwards.

Clockwise moment = anticlockwise moment.

There are three obvious choices, A, B and C, for the point to take moments about. If you choose A or C you will get an equation in just one unknown because one force acts through your point. Whichever point you choose, you will get the same final answer, so choose the option that makes the algebra as simple as possible.

Substituting the value of R_C into the first equation.

Example 8

A uniform beam AB, of mass 40 kg and
length 5 m, rests horizontally on supports
at C and D, where $AC = DB = 1$ m.
When a man of mass 80 kg stands

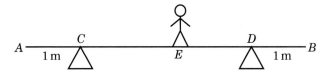

on the beam at E the magnitude of the reaction at D is twice the magnitude of the reaction at C.
By modelling the beam as a rod and the man as a particle, find the distance AE.

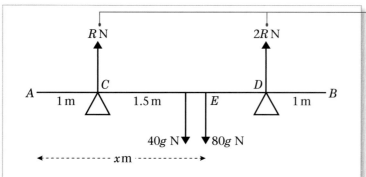

Draw a diagram showing the forces.

Because you are told a relationship between the reaction at C and the reaction at D you can use this on your diagram.

Resolving vertically:

$R + 2R = 40g + 80g = 120g$

$\Rightarrow \quad R = 40g = 392$

The rod is in equilibrium so there is no resultant force.

Let the distance AE be x m.

Taking moments about A:

$40g \times 2.5 + 80g \times x = 40g \times 1 + 80g \times 4$

$100g + 80g \times x = 40g + 320g = 360g$

$80g \times x = 360g - 100g = 260g$

$\Rightarrow \quad x = \dfrac{260g}{80g} = \dfrac{26}{8} = 3.25$

The man stands 3.25 m from A.

Clockwise moment = anticlockwise moment.

How have you used the modelling assumptions in the question?

- Since the beam is a rod, it remains straight.
- Since the man is a particle, his weight acts at the point E.

Example 9

A uniform rod AB of length 4 m and mass 12 kg is resting in a horizontal position on supports at C and D, with $AC = DB = 0.5$ m. When a particle of mass m kg is placed on a rod at point B the rod is on the point of turning about D. Find the value of m.

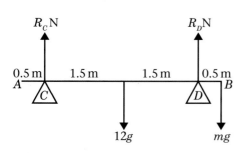

Draw a diagram showing all the forces.

If the rod is about to turn about D then there is no contact force at C.

If the rod is on the point of turning about D then $R_C = 0$.

Taking moments about D: $mg \times 0.5 = 12g \times 1.5$

$m \times 0.5 = 12 \times 1.5$

$m = 12 \times 3$

$= 36$

The rod is in equilibrium, so clockwise moment = anticlockwise moment.

You could choose to take moments about any point on the rod but by choosing D you do not need to find the value of R_D to solve the problem.

Exercise 5C

1 *AB* is a uniform rod of length 5 m and weight 20 N. In these diagrams *AB* is resting in a horizontal position on supports at *C* and *D*. In each case, find the magnitudes of the reactions at *C* and *D*.

a A —1 m— C —————3 m————— D —1 m— B

b A ————2 m———— C ————2 m———— D —1 m— B

c A ——1.5 m—— C ———2.5 m——— D —1 m— B

d A ——1.5 m—— C ————2.7 m———— D —0.8 m— B

2 Each of these diagrams shows a light rod in equilibrium in a horizontal position under the action of a set of forces. Find the value of the unknown forces and distances.

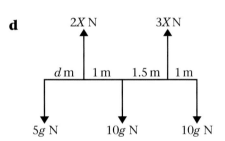

a X N ↑ — 1 m — ↑ Y N — 2 m — 1 m ; 10 N ↓ , 15 N ↓

b 15 N ↑ — 1 m — ↑ Y N — 2 m — 1 m ; 20 N ↓ , X N ↓

c $5g$ N ↑ — 2 m — X N ↑ 2 m | 3 m | d m ; $5g$ N ↓ , $10g$ N ↓ , $15g$ N ↓

d $2X$ N ↑ d m | 1 m — $3X$ N ↑ 1.5 m | 1 m ; $5g$ N ↓ , $10g$ N ↓ , $10g$ N ↓

3 Jack and Jill are playing on a see-saw made from a uniform plank *AB* of length 5 m pivoted at *M*, the mid-point of *AB*. Jack has mass 35 kg and Jill has mass 28 kg. Jill sits at *A*. Where must Jack sit for the plank to be in equilibrium when horizontal?

4 A uniform rod *AB* of length 3 m and mass 12 kg is pivoted at *C*, where *AC* = 1 m. Calculate the vertical force that must be applied at *A* to maintain equilibrium with the rod horizontal.

5 A broom consists of a broomstick of length 130 cm and mass 5 kg and a broomhead of mass 5.5 kg attached at one end. By modelling the broomstick as a rod and the broomhead as a particle, find where a support should be placed so that the broom will balance horizontally.

6 A uniform rod *AB* of length 4 m and weight 20 N is suspended horizontally by two vertical strings attached at *A* and at *B*. A particle of weight 10 N is attached to the rod at point *C*, where *AC* = 1.5 m. Find the magnitudes of the tensions in the two strings.

7 A uniform plank AB of length 5 m and mass 30 kg is resting horizontally on supports at C and D, where $AC = 1$ m and $AD = 3.5$ m. When a particle of mass 14 kg is attached to the rod at point E the magnitude of the reaction at C is equal to the magnitude of the reaction at D. Find the distance AE.

8 A uniform rod AB has length 4 m and mass 8 kg. It is resting in a horizontal position on supports at points C and D where $AC = 1$ m and $AD = 2.5$ m. A particle of mass m kg is placed at point E where $AE = 3.3$ m. Given that the rod is about to tilt about D, calculate the value of m.

9 A uniform bar AB of length 6 m and weight 40 N is resting in a horizontal position on supports at points C and D where $AC = 2$ m and $AD = 5$ m. When a particle of weight 30 N is attached to the bar at point E the bar is on the point of tilting about C. Calculate the distance AE.

10 A plank AB of mass 12 kg and length 3 m is in equilibrium in a horizontal position resting on supports at C and D where $AC = 0.7$ m and $DB = 1.1$ m. A boy of mass 32 kg stands on the plank at point E. The plank is about to tilt about D. By modelling the plank as a uniform rod and the boy as a particle, calculate the distance AE.

11 A uniform rod AB has length 5 m and weight 20 N. The rod is resting on supports at points C and D where $AC = 2$ m and $BD = 1$ m.

  a Find the magnitudes of the reactions at C and D.

  A particle of weight 12 N is placed on the rod at point A.

  b Show that this causes the rod to tilt about C.

  A second particle of weight 12 N is placed on the rod at E to hold it in equilibrium.

  c How far must E be from A?

5.4 You can solve problems about non-uniform bodies by finding or using the centre of mass.

■ **The mass of a non-uniform rigid body can be modelled as acting at its centre of mass.**

If a rigid body is not uniform then you will either be told where the centre of mass lies, or you will be asked to find out where it is.

Example 10

Sam and Tamsin are sitting on a non-uniform plank AB of mass 25 kg and length 4 m. The plank is pivoted at M, the mid-point of AB. The centre of mass of AB is at C where AC is 1.8 m. Sam has mass 35 kg.
Tamsin has mass 25 kg and sits at A.
Where must Sam sit for the plank to be horizontal?

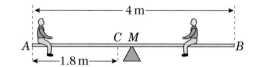

Modelling the plank as a rod and the two children as particles:

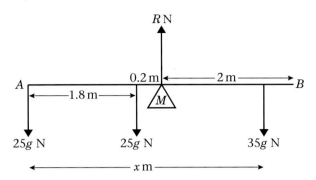

Draw a diagram to represent the situation.

Suppose that Sam sits at point x m from A.

Taking moments about M:

$25g \times 2 + 25g \times 0.2 = 35g \times (x - 2)$

$\Rightarrow \quad 50 + 5 = 35x - 70$

$\qquad 35x = 55 + 70$

$\qquad \qquad = 125$

$\Rightarrow \qquad x = 3.57$

Sam should sit 3.57 m from end A (or 0.43 m from end B).

There is a reaction at M but we do not know the magnitude of this reaction and we are not asked to find it, so M is a good choice for the point to take moments about.

Dividing through by the common factor g.

Example 11

A rod AB is 3 m long and has weight 20 N. It is in a horizontal position resting on supports at points C and D, where $AC = 1$ m and $AD = 2.5$ m. The magnitude of the reaction at C is three times the magnitude of the reaction at D. Find the distance of the centre of mass of the rod from A.

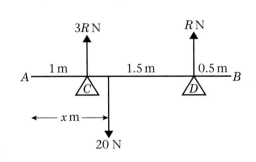

Draw a diagram. Make sure that you have used all the information given in the question.

Suppose that the centre of mass acts at a point x m from A.

Resolving vertically, $3R + R = 20$

$\qquad \qquad \qquad \quad R = 5$

Whichever point you choose to take moments about, you are going to need to know the magnitude of R.

Taking moments about A:

$20 \times x = 15 \times 1 + 5 \times 2.5$

$\qquad = 15 + 12.5$

$20x = 27.5$

$\qquad x = 1.38 \text{ (3 s.f.)}$

Using your value of R.

The centre of mass is 1.38 m from A, to 3 s.f.

Exercise 5D

1 A non-uniform rod AB of length 4 m and weight 6 N rests horizontally on two supports at A and B. Given that the centre of mass of the rod is 2.4 m from the end A, find the reactions at the two supports.

2 A non-uniform bar AB of length 5 m is supported horizontally on supports at A and B. The reactions at these supports are $3g$ N and $7g$ N respectively. Find the position of the centre of mass.

3 A non-uniform plank AB of length 4 m and weight 120 N is pivoted at its mid-point. The plank is in equilibrium in a horizontal position with a child of weight 200 N sitting at A and a child of weight 300 N sitting at B. By modelling the plank as a rod and the two children as particles find the distance of the centre of mass of the plank from A.

4 A non-uniform rod AB of length 5 m and mass 15 kg rests horizontally suspended from the ceiling by two vertical strings attached to C and D, where $AC = 1$ m and $AD = 3.5$ m.

    a Given that the centre of mass is at E where $AE = 3$ m, find the magnitudes of the tensions in the strings.

    When a particle of mass 10 kg is attached to the rod at F the rod is just about to rotate about D.

    b Find the distance AF.

Mixed exercise 5E

1

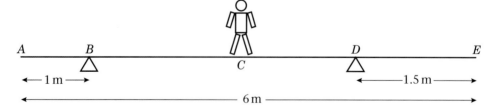

A plank AE, of length 6 m and weight 100 N, rests in a horizontal position on supports at B and D, where $AB = 1$ m and $DE = 1.5$ m. A child of weight 145 N stands at C, the mid-point of AE, as shown in the diagram above. The child is modelled as a particle and the plank as a uniform rod. The child and the plank are in equilibrium. Calculate

    a the magnitude of the force exerted by the support on the plank at B,

    b the magnitude of the force exerted by the support on the plank at D.

The child now stands at a different point F on the plank. The plank is in equilibrium and on the point of tilting about D.

c Calculate the distance DF.

2

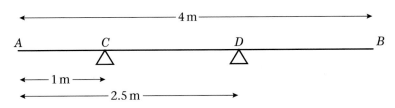

A uniform rod AB has length 4 m and weight 150 N. The rod rests in equilibrium in a horizontal position, smoothly supported at points C and D, where AC = 1 m and AD = 2.5 m as shown in the diagram above. A particle of weight W N is attached to the rod at a point E where AE = x metres. The rod remains in equilibrium and the magnitude of the reaction at C is now equal to the magnitude of the reaction at D.

a Show that $W = \dfrac{150}{7 - 4x}$

b Hence deduce the range of possible values of x.

3

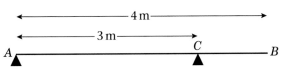

A uniform plank AB has mass 40 kg and length 4 m. It is supported in a horizontal position by two smooth pivots. One pivot is at the end A and the other is at the point C where AC = 3 m, as shown in the diagram above. A man of mass 80 kg stands on the plank which remains in equilibrium. The magnitude of the reaction at A is twice the magnitude of the reaction at C. The magnitude of the reaction at C is R N. The plank is modelled as a rod and the man is modelled as a particle.

a Find the value of R.

b Find the distance of the man from A.

c State how you have used the modelling assumption that
   i the plank is uniform,
   ii the plank is a rod,
   iii the man is a particle.

4

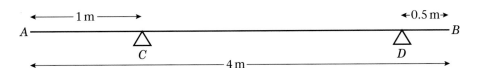

A non-uniform rod AB has length 4 m and weight 150 N. The rod rests horizontally in equilibrium on two smooth supports C and D, where AC = 1 m and DB = 0.5 m, as shown in the diagram above. The centre of mass of AB is x metres from A. A particle of weight W N is placed on the rod at A. The rod remains in equilibrium and the magnitude of the reaction of C on the rod is 100 N.

a Show that $550 + 7W = 300x$.

The particle is now removed from A and placed on the rod at B. The rod remains in equilibrium and the reaction of C on the rod now has magnitude 52 N.

b Obtain another equation connecting W and x.

c Calculate the value of x and the value of W.

5

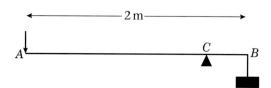

A lever consists of a uniform steel rod AB of weight 100 N and length 2 m, which rests on a small smooth pivot at a point C. A load of weight 1700 N is suspended from the end B of the rod by a rope. The lever is held in equilibrium in a horizontal position by a vertical force applied at the end A, as shown in the diagram above. The rope is modelled as a light string.

a Given that $BC = 0.25$ m find the magnitude of the force applied at A.

The position of the pivot is changed so that the rod remains in equilibrium when the force at A has magnitude 150 N.

b Find, to the nearest centimetre, the new distance of the pivot from B.

6

A plank AB has length 4 m. It lies on a horizontal platform, with the end A lying on the platform and the end B projecting over the edge, as shown above. The edge of the platform is at the point C.

Jack and Jill are experimenting with the plank. Jack has mass 48 kg and Jill has mass 36 kg. They discover that if Jack stands at B and Jill stands at A and $BC = 1.8$ m, the plank is in equilibrium and on the point of tilting about C.

a By modelling the plank as a uniform rod, and Jack and Jill as particles, find the mass of the plank.

They now alter the position of the plank in relation to the platform so that, when Jill stands at B and Jack stands at A, the plank is again in equilibrium and on the point of tilting about C.

b Find the distance BC in this position.

7

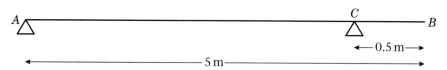

A plank of wood AB has mass 12 kg and length 5 m. It rests in a horizontal position on two smooth supports. One support is at the end A. The other is at the point C, 0.5 m from B, as shown in the diagram above. A girl of mass 30 kg stands at B with the plank in equilibrium.

a By modelling the plank as a uniform rod and the girl as a particle, find the reaction on the plank at A.

The girl gets off the plank. A boulder of mass m kg is placed on the plank at A and a man of mass 93 kg stands on the plank at B. The plank remains in equilibrium and is on the point of tilting about C.

b By modelling the plank again as a uniform rod, and the man and the boulder as particles, find the value of m.

8

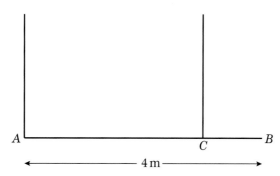

A plank AB has mass 50 kg and length 4 m. A load of mass 25 kg is attached to the plank at B. The loaded plank is held in equilibrium, with AB horizontal, by two vertical ropes attached at A and C, as shown in the diagram. The plank is modelled as a uniform rod and the load as a particle. Given that the tension in the rope at C is four times the tension in the rope at A, calculate

a the tension in the rope at C,

b the distance CB.

9

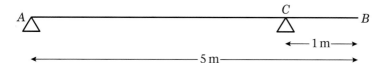

A uniform beam AB has weight 200 N and length 5 m. The beam rests in equilibrium in a horizontal position on two smooth supports. One support is at end A and the other is at a point C on the beam, where $BC = 1$ m, as shown in the diagram. The beam is modelled as a uniform rod.

a Find the reaction on the beam at C.

A woman of weight 500 N stands on the beam at the point D. The beam remains in equilibrium. The reactions on the beam at A and C are now equal.

b Find the distance AD.

Summary of key points

1 A force applied to a rigid body can cause the body to rotate.

2 The moment of a force measures the turning effect of the force on the body on which it acts.

3 The moment of the force F about a point P is the product of the magnitude of the force F and the perpendicular distance from the point P to the line of action of F.

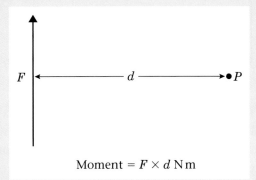

Moment $= F \times d$ N m

4 When you have several coplanar forces acting on a body, you can add the moments about a point. You need to choose a positive direction (clockwise or anticlockwise) and consider the sense of rotation of each moment.

5 When the distance given is not the perpendicular distance of the line of action from the pivot you can find the moment by resolving the force into components.

6 If a body is resting in equilibrium the resultant force in any direction is zero, and the sum of the moments about any point is zero.

7 The mass of a non-uniform rigid body can be modelled as acting at its centre of mass.

After completing this chapter you should be able to

- use vectors to describe the forces acting on a particle
- give the displacement, velocity or acceleration of a particle as a vector.

Vectors

Vectors can provide an alternative, often simpler, method of solving problems in mechanics. In everyday life vectors are used in many situations, especially where a problem involves movement in three dimensions.

Computer graphics software uses vectors to model the movements of animated characters.

Engineers need to consider forces acting in three dimensions when they design buildings, bridges and other structures.

An air traffic control centre follows the course of incoming and outgoing flights. Their paths are recorded as vectors.

6.1 You can use vectors to describe displacements.

■ **A vector is a quantity which has both magnitude and direction.**

Examples of vectors:

- the velocity of an object – 'moving due north at $20\,\mathrm{m\,s^{-1}}$,
- a force applied to an object – 'a horizontal force of $7\,\mathrm{N}$',
- displacement – '5 m to the left'.

Example 1

A girl walks 2 km due east from a fixed point O to A, and then 3 km due south from A to B. Find the total distance walked, and describe the displacement of B from O.

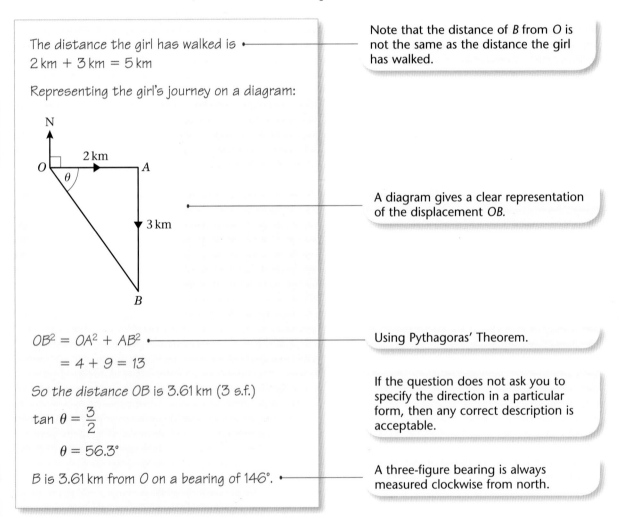

The distance the girl has walked is
$2\,\mathrm{km} + 3\,\mathrm{km} = 5\,\mathrm{km}$

Note that the distance of B from O is not the same as the distance the girl has walked.

Representing the girl's journey on a diagram:

A diagram gives a clear representation of the displacement OB.

$OB^2 = OA^2 + AB^2$

Using Pythagoras' Theorem.

$\quad = 4 + 9 = 13$

So the distance OB is 3.61 km (3 s.f.)

$\tan \theta = \dfrac{3}{2}$

$\quad \theta = 56.3°$

If the question does not ask you to specify the direction in a particular form, then any correct description is acceptable.

B is 3.61 km from O on a bearing of 146°.

A three-figure bearing is always measured clockwise from north.

Example 2

In an orienteering exercise, a cadet leaves the starting point S and walks 15 km on a bearing of 120° to reach A, the first checkpoint. From A he walks 9 km on a bearing of 240° to the second checkpoint, at B. From B he returns directly to S. Describe the displacement of S from B.

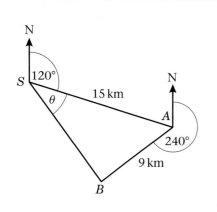

$SB^2 = 15^2 + 9^2 - 2 \times 15 \times 9 \times \cos 60$

$\quad = 171$

$SB = 13.1 \text{ km (3 s.f.)}$

$\dfrac{\sin \theta}{9} = \dfrac{\sin 60}{\sqrt{171}}$

$\sin \theta = \dfrac{9 \times \sin 60}{\sqrt{171}} = 0.596 \ldots$

$\theta = 36.6°$

S is 13.1 km from B on a bearing of 337°.

Start by drawing a diagram

$\angle SAB = 360° - (240° + 60°) = 60°$

Use the Cosine Rule in triangle SAB.

Use the Sine Rule.

You are asked to describe the journey from B to S, so you need to measure the bearing of S from B.

The bearing from B from S is 157°, so the required angle is $360° - (180° - 157°)$.

Exercise 6A

1 A bird flies 5 km due north and then 7 km due east. How far is the bird from its original position, and in what direction?

2 A girl cycles 4 km due west then 6 km due north. Calculate the total distance she has cycled and her displacement from her starting point.

3 A man walks 3 km due east and then 5 km northeast. Find his distance and bearing from his original position.

4 In an orienteering exercise, a team hikes 8 km from the starting point, S, on a bearing of 300° then 6 km on a bearing 040° to the finishing point, F. Find the magnitude and direction of the displacement from S to F.

5 A boat travels 20 km on a bearing of 060°, followed by 15 km on a bearing of 110°. What course should it take to return to its starting point by the shortest route?

6 An aeroplane flies from airport A to airport B 80 km away on a bearing of 070°. From B the aeroplane flies to airport C, 60 km from B. Airport C is 90 km from A. Find the two possible directions for the course set by the aeroplane on the second stage of its journey.

7 In a regatta, a yacht starts at point O, sails 2 km due east to A, 3 km due south from A to B, and then 4 km on a bearing of 280° from B to C. Find the displacement vector of C from O.

6.2 You can add vectors, and represent vectors using line segments.

■ **A vector can be represented as a directed line segment.**

The direction of the line is the direction of the vector,
and the length of the line represents the magnitude of the vector.

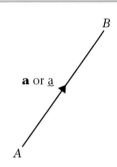

a or $\underline{a}$

The vector describing the displacement of B from A is usually
written as **AB**, $\overrightarrow{AB}$ or $\overline{AB}$. If a single letter is used to represent a
vector it is usually in bold or underlined.

■ **Two vectors are equal if and only if they have the same
magnitude and the same direction.**

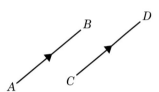

■ **Two vectors are parallel if and only if they have the
same direction.**

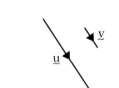

■ **You can add vectors using the triangle law of addition.**
$$\overrightarrow{AC} = \overrightarrow{AB} + \overrightarrow{BC}$$

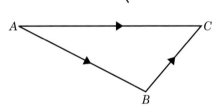

Example 3

$OACB$ is a parallelogram. The points P, Q, M
and N are the mid-points of the sides.
$\overrightarrow{OA} = \mathbf{a}$ and $\overrightarrow{OB} = \mathbf{b}$

a Express $\overrightarrow{OC}, \overrightarrow{AB}, \overrightarrow{QC}, \overrightarrow{CN}$ and $\overrightarrow{QN}$ in terms
of **a** and **b**.

b What can you deduce about $\overrightarrow{QN}$?

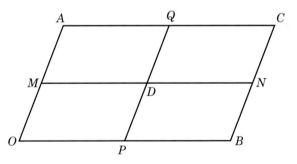

a $\overrightarrow{OC} = \overrightarrow{OA} + \overrightarrow{AC}$ •————— Use the triangle rule for addition of vectors.

   $= \overrightarrow{OA} + \overrightarrow{OB}$ •—————

   $= \mathbf{a} + \mathbf{b}$

Opposite sides of a parallelogram are parallel and equal in lengths, so $\overrightarrow{OB} = \overrightarrow{AC}$.

  $\overrightarrow{AB} = \overrightarrow{AO} + \overrightarrow{OB}$ •—————

    $= -\overrightarrow{OA} + \overrightarrow{OB}$ •—————

    $= -\mathbf{a} + \mathbf{b}, \quad \text{or} \quad \mathbf{b} - \mathbf{a}$

Using the triangle rule again.

$\overrightarrow{AO}$ is equal in length to $\overrightarrow{OA}$, but in the opposite direction.

$$\overrightarrow{QC} = \overrightarrow{OP} = \tfrac{1}{2}\overrightarrow{OB} = \tfrac{1}{2}\mathbf{b}$$

$$\overrightarrow{CN} = \tfrac{1}{2}\overrightarrow{CB} = \tfrac{1}{2}\overrightarrow{AO} = \tfrac{1}{2}(-\mathbf{a}) = -\tfrac{1}{2}\mathbf{a}$$

$$\overrightarrow{QN} = \overrightarrow{QC} + \overrightarrow{CN} = \tfrac{1}{2}\mathbf{b} - \tfrac{1}{2}\mathbf{a}$$

$$= \tfrac{1}{2}(\mathbf{b} - \mathbf{a})$$

b $\quad \overrightarrow{QN} = \tfrac{1}{2}(\mathbf{b} - \mathbf{a}) = \tfrac{1}{2}\overrightarrow{AB}$, so QN is

parallel to AB, and half as long.

OACB is a parallelogram, so $\overrightarrow{OB} = \overrightarrow{AC}$.
P and *Q* are the mid-points of sides *OB* and *AC* respectively,
so $\overrightarrow{OP} = \overrightarrow{PB} = \overrightarrow{AQ} = \overrightarrow{QC} = \tfrac{1}{2}\mathbf{b}$.

Similarly, you can use the other pair of parallel sides to deduce *CN*.

Use the triangle rule.

If you can express one vector as a scalar multiple of the other then the two vectors must be parallel since they have the same direction.

Example 4

In triangle *OAB*, *M* is the mid-point of *OA* and *N* divides *AB* in the ratio $1:2$.
$\overrightarrow{OM} = \mathbf{a}$, and $\overrightarrow{OB} = \mathbf{b}$
Express $\overrightarrow{ON}$ in terms of **a** and **b**.

$$\overrightarrow{ON} = \overrightarrow{OA} + \overrightarrow{AN}$$

$$\overrightarrow{OA} = 2\mathbf{a}$$

$$\overrightarrow{AN} = \tfrac{1}{3}\overrightarrow{AB}$$

but $\quad \overrightarrow{AB} = \overrightarrow{AO} + \overrightarrow{OB} = -2\mathbf{a} + \mathbf{b}$

so $\quad \overrightarrow{AN} = \tfrac{1}{3}(-2\mathbf{a} + \mathbf{b})$

and $\quad \overrightarrow{ON} = 2\mathbf{a} + \tfrac{1}{3}(-2\mathbf{a} + \mathbf{b})$

$$= 2\mathbf{a} - \tfrac{2}{3}\mathbf{a} + \tfrac{1}{3}\mathbf{b}$$

$$= 1\tfrac{1}{3}\mathbf{a} + \tfrac{1}{3}\mathbf{b}$$

Use the triangle law of addition to find another way of expressing $\overrightarrow{ON}$.

You also need to use the addition law to rewrite $\overrightarrow{AB}$. Each part of the expression needs to be written in terms of **a** and **b**

Substitute your results and simplify the expression.

Example 5

OABC is a parallelogram. *P* is the point where the diagonals *OB* and *AC* intersect.

The vectors **a** and **c** are equal to $\overrightarrow{OA}$ and $\overrightarrow{OC}$ respectively.

Prove that the diagonals bisect each other.

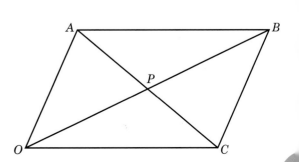

If the diagonals bisect each other, then P must be the mid-point of OB and the mid-point of AC.

From the diagram,

$$\overrightarrow{OB} = \overrightarrow{OC} + \overrightarrow{CB} = \mathbf{c} + \mathbf{a}$$

and

$$\overrightarrow{AC} = \overrightarrow{AO} + \overrightarrow{OC}$$

$$= -\overrightarrow{OA} + \overrightarrow{OC} = -\mathbf{a} + \mathbf{c}$$

> Express $\overrightarrow{OB}$ and $\overrightarrow{AC}$ in terms of $\mathbf{a}$ and $\mathbf{c}$.

P lies on $OB \Rightarrow \overrightarrow{OP} = \lambda(\mathbf{c} + \mathbf{a})$

P lies on $AC \Rightarrow \overrightarrow{OP} = \overrightarrow{OA} + \overrightarrow{AP}$

$$= \mathbf{a} + \mu(-\mathbf{a} + \mathbf{c})$$

> Use the fact that P lies on both diagonals to find two different routes from O to P, giving two different forms of $\overrightarrow{OP}$.

$$\Rightarrow \quad \lambda(\mathbf{c} + \mathbf{a}) = \mathbf{a} + \mu(-\mathbf{a} + \mathbf{c})$$

> The two expressions for $\overrightarrow{OP}$ must be equal.

$$\Rightarrow \quad \lambda = 1 - \mu \quad \text{and} \quad \lambda = \mu$$

> Form and solve a pair of simultaneous equations (by equating the coefficients of $\mathbf{a}$ and $\mathbf{c}$).

$\Rightarrow \quad \lambda = \mu = \dfrac{1}{2}$, so P is the mid-point of both diagonals, so the diagonals bisect each other.

> If P is halfway along the line then it must be the mid-point.

Exercise 6B

1

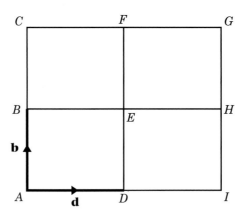

$ACGI$ is a square, B is the mid-point of AC, F is the mid-point of CG, H is the mid-point of GI, and D is the mid-point of AI.

Vectors $\mathbf{b}$ and $\mathbf{d}$ are represented in magnitude and direction by $\overrightarrow{AB}$ and $\overrightarrow{AD}$ respectively. Find, in terms of $\mathbf{b}$ and $\mathbf{d}$, the vectors represented in magnitude and direction by

a $\overrightarrow{AC}$, **b** $\overrightarrow{BE}$, **c** $\overrightarrow{HG}$, **d** $\overrightarrow{DF}$,

e $\overrightarrow{AE}$, **f** $\overrightarrow{DH}$, **g** $\overrightarrow{HB}$, **h** $\overrightarrow{FE}$,

i $\overrightarrow{AH}$, **j** $\overrightarrow{BI}$, **k** $\overrightarrow{EI}$, **l** $\overrightarrow{FB}$.

2

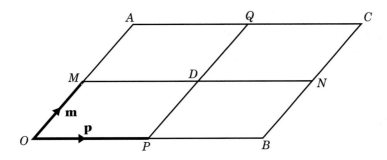

OACB is a parallelogram. *M, Q, N* and *P* are the mid-points of *OA, AC, BC* and *OB* respectively.

Vectors **p** and **m** are equal to $\overrightarrow{OP}$ and $\overrightarrow{OM}$ respectively. Express in terms of **p** and **m**

a $\overrightarrow{OA}$	**b** $\overrightarrow{OB}$	**c** $\overrightarrow{BN}$	**d** $\overrightarrow{DQ}$
e $\overrightarrow{OD}$	**f** $\overrightarrow{MQ}$	**g** $\overrightarrow{OQ}$	**h** $\overrightarrow{AD}$
i $\overrightarrow{CD}$	**j** $\overrightarrow{AP}$	**k** $\overrightarrow{BM}$	**l** $\overrightarrow{NO}$.

3

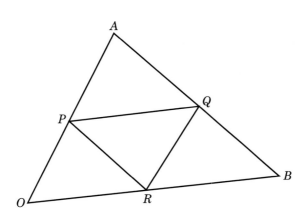

OAB is a triangle. *P, Q* and *R* are the mid-points of *OA, AB* and *OB* respectively. $\overrightarrow{OP}$ and $\overrightarrow{OR}$ are equal to **p** and **r** respectively. Find, in terms of **p** and **r**

a $\overrightarrow{OA}$	**b** $\overrightarrow{OB}$	**c** $\overrightarrow{AB}$	**d** $\overrightarrow{AQ}$
e $\overrightarrow{OQ}$	**f** $\overrightarrow{PQ}$	**g** $\overrightarrow{QR}$	**h** $\overrightarrow{BP}$.

Use parts **b** and **f** to prove that triangle *PAQ* is similar to triangle *OAB*.

4

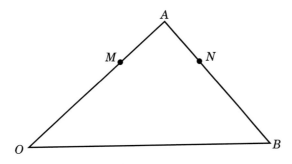

OAB is a triangle. $\overrightarrow{OA} = \mathbf{a}$ and $\overrightarrow{OB} = \mathbf{b}$. The point *M* divides *OA* in the ratio $2:1$. *MN* is parallel to *OB*. Express the vector $\overrightarrow{ON}$ in terms of **a** and **b**.

5

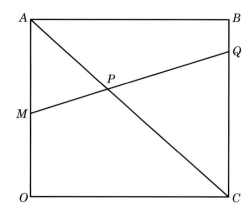

OABC is a square. M is the mid-point of OA, and Q divides BC in the ratio 1:3. AP and MQ meet at P. If $\overrightarrow{OA}$ = **a** and $\overrightarrow{OC}$ = **c**, express $\overrightarrow{OP}$ in terms of **a** and **c**.

6.3 You can describe vectors using i, j notation.

■ **A unit vector is a vector of length 1. The unit vectors along the Cartesian axes are usually denoted by i and j respectively.**

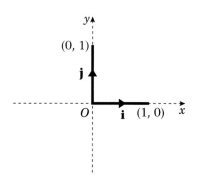

■ **You can write any two-dimensional vector in the form ai + bj**

By the triangle law of addition:
$\overrightarrow{AC} = \overrightarrow{AB} + \overrightarrow{BC}$
$= 5\mathbf{i} + 2\mathbf{j}$

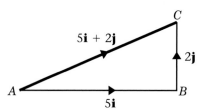

Example 6

Draw a diagram to represent the vector $-3\mathbf{i} + \mathbf{j}$.

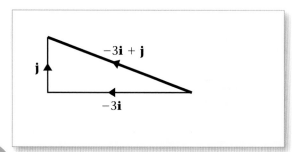

3 units in the direction of the unit vector $-\mathbf{i}$ and 1 unit in the direction of the unit vector **j**.

Exercise 6C

Express the vectors $\mathbf{v}_1$, $\mathbf{v}_2$, $\mathbf{v}_3$, $\mathbf{v}_4$, $\mathbf{v}_5$ and $\mathbf{v}_6$ using the $\mathbf{i}$, $\mathbf{j}$ notation.

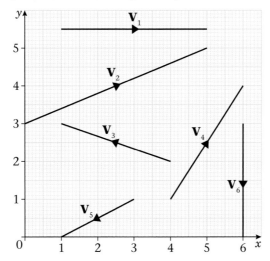

6.4 **You can solve problems with vectors written using the i, j notation.**

■ **When vectors are written in terms of the unit vectors i and j you can add them together by adding the terms in i and j separately. You subtract vectors in a similar way.**

Example 7

Given $\mathbf{p} = 2\mathbf{i} + 3\mathbf{j}$ and $\mathbf{q} = 5\mathbf{i} + \mathbf{j}$, find $\mathbf{p} + \mathbf{q}$ in terms of $\mathbf{i}$ and $\mathbf{j}$.

$$\mathbf{p} + \mathbf{q} = (2\mathbf{i} + 3\mathbf{j}) + (5\mathbf{i} + \mathbf{j})$$
$$= (2\mathbf{i} + 5\mathbf{i}) + (3\mathbf{j} + \mathbf{j})$$
$$= 7\mathbf{i} + 4\mathbf{j}$$

Rearrange with the **i** terms together and the **j** terms together.

Simplify the answer.

Example 8

Given $\mathbf{a} = 5\mathbf{i} + 2\mathbf{j}$ and $\mathbf{b} = 3\mathbf{i} - 4\mathbf{j}$, find $2\mathbf{a} - \mathbf{b}$ in terms of $\mathbf{i}$ and $\mathbf{j}$.

$$2\mathbf{a} = 2(5\mathbf{i} + 2\mathbf{j})$$
$$= 10\mathbf{i} + 4\mathbf{j}$$

When you double **a** the **i** term and the **j** term are both doubled – just the same as when you multiply out a bracket in algebra.

$$2\mathbf{a} - \mathbf{b} = (10\mathbf{i} + 4\mathbf{j}) - (3\mathbf{i} - 4\mathbf{j})$$
$$= (10\mathbf{i} - 3\mathbf{i}) + (4\mathbf{j} - (-4\mathbf{j}))$$
$$= 7\mathbf{i} + 8\mathbf{j}$$

Rearrange with the **i** terms and the **j** terms together.

Take care when subtracting a negative term.

■ When a vector is given in terms of the unit vectors **i**, and **j** you can find its magnitude using Pythagoras' Theorem. The magnitude of a vector **a** is written |**a**|.

Example 9

Given that **v** = 3**i** − 7**j**, find the magnitude of **v**.

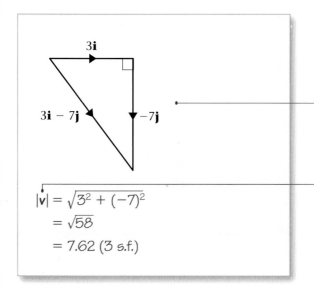

Because **i** and **j** are perpendicular vectors, you have a right-angled triangle.

$$|v| = \sqrt{3^2 + (-7)^2}$$
$$= \sqrt{58}$$
$$= 7.62 \ (3 \ \text{s.f.})$$

|**v**| means the magnitude of **v**.

Example 10

Find the angle between the vector 4**i** + 5**j** and the positive *x*-axis.

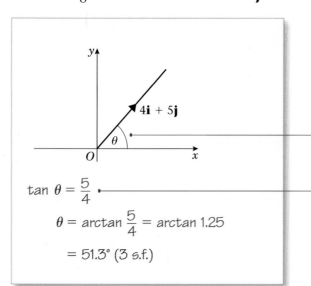

Identify the angle that you need to find. A diagram always helps.

$$\tan \theta = \frac{5}{4}$$

You have a right-angled triangle, so finding the angle is straight forward.

$$\theta = \arctan \frac{5}{4} = \arctan 1.25$$
$$= 51.3° \ (3 \ \text{s.f.})$$

Exercise 6D

1 Given that **a** = 2**i** + 3**j** and **b** = 4**i** − **j**, find these terms of **i** and **j**.

   a **a** + **b**   **b** 3**a** + **b**   **c** 2**a** − **b**   **d** 2**b** + **a**

   e 3**a** − 2**b**   **f** **b** − 3**a**   **g** 4**b** − **a**   **h** 2**a** − 3**b**

2 Find the magnitude of each of these vectors.

   a $3\mathbf{i} + 4\mathbf{j}$ **b** $6\mathbf{i} - 8\mathbf{j}$ **c** $5\mathbf{i} + 12\mathbf{j}$ **d** $2\mathbf{i} + 4\mathbf{j}$

   e $3\mathbf{i} - 5\mathbf{j}$ **f** $4\mathbf{i} + 7\mathbf{j}$ **g** $-3\mathbf{i} + 5\mathbf{j}$ **h** $-4\mathbf{i} - \mathbf{j}$

3 Find the angle that each of these vectors makes with the positive x-axis.

   a $3\mathbf{i} + 4\mathbf{j}$ **b** $6\mathbf{i} - 8\mathbf{j}$ **c** $5\mathbf{i} + 12\mathbf{j}$ **d** $2\mathbf{i} + 4\mathbf{j}$

4 Find the angle that each of these vectors makes with the positive y-axis.

   a $3\mathbf{i} - 5\mathbf{j}$ **b** $4\mathbf{i} + 7\mathbf{j}$ **c** $-3\mathbf{i} + 5\mathbf{j}$ **d** $-4\mathbf{i} - \mathbf{j}$

5 Given that $\mathbf{a} = 2\mathbf{i} + 5\mathbf{j}$ and $\mathbf{b} = 3\mathbf{i} - \mathbf{j}$, find

   a λ if $\mathbf{a} + \lambda\mathbf{b}$ is parallel to the vector $\mathbf{i}$,   **b** μ if $\mu\mathbf{a} + \mathbf{b}$ is parallel to the vector $\mathbf{j}$,

6 Given that $\mathbf{c} = 3\mathbf{i} + 4\mathbf{j}$ and $\mathbf{d} = \mathbf{i} - 2\mathbf{j}$, find

   a λ if $\mathbf{c} + \lambda\mathbf{d}$ is parallel to $\mathbf{i} + \mathbf{j}$,   **b** μ if $\mu\mathbf{c} + \mathbf{d}$ is parallel to $\mathbf{i} + 3\mathbf{j}$,

   c s if $\mathbf{c} - s\mathbf{d}$ is parallel to $2\mathbf{i} + \mathbf{j}$,   **d** t if $\mathbf{d} - t\mathbf{c}$ is parallel to $-2\mathbf{i} + 3\mathbf{j}$.

7 In this question, the horizontal unit vectors $\mathbf{i}$ and $\mathbf{j}$ are directed due east and due north respectively.

   Find the magnitude and bearing of these vectors.

   a $2\mathbf{i} + 3\mathbf{j}$ **b** $4\mathbf{i} - \mathbf{j}$ **c** $-3\mathbf{i} + 2\mathbf{j}$ **d** $-2\mathbf{i} - \mathbf{j}$

6.5 You can express the velocity of a particle as a vector.

■ **The velocity of a particle is a vector in the direction of motion. Its magnitude is the speed of the particle. The velocity is usually denoted by v.**

If a particle is moving with constant velocity $\mathbf{v}$ m s^{-1}, then after time t seconds it will have moved $\mathbf{v}t$ m. The displacement is parallel to the velocity. The magnitude of the displacement is the distance from the starting point.

Example 11

A particle is moving with constant velocity $\mathbf{v} = (3\mathbf{i} + \mathbf{j})$ m s^{-1}. Find **a** the speed of the particle, **b** the distance moved every 4 seconds.

   a $\mathbf{v} = (3\mathbf{i} + \mathbf{j})$,

         so speed $|\mathbf{v}| = \sqrt{3^2 + 1^2} = \sqrt{10} = 3.16$ m s^{-1} The speed is the magnitude of the velocity.

b   **Method 1**

$$\text{Displacement} = \mathbf{v}t = (3\mathbf{i} + \mathbf{j}) \times 4$$

$$= (12\mathbf{i} + 4\mathbf{j})$$

Therefore distance moved is given by

$$|\mathbf{v}t| = \sqrt{12^2 + 4^2} = \sqrt{144 + 16} = \sqrt{160}$$

$$= 12.6\,\text{m}$$

Method 2

$$\mathbf{v} = (3\mathbf{i} + \mathbf{j}),$$

$$\text{so speed} = |\mathbf{v}| = \sqrt{3^2 + 1^2} = \sqrt{10} = 3.16\,\text{m s}^{-1}$$

$$\text{Distance} = 3.162 \times 4 = 12.6\,\text{m}$$

> Find the displacement of the particle in 4 seconds.

> Displacement = $\mathbf{v} \times t$

> Distance is magnitude of displacement.

> Find the speed of the particle.

> Distance = speed × time
> Use the unrounded value in your calculation.

Exercise 6E

1 Find the speed of a particle moving with these velocities:

  a $3\mathbf{i} + 4\mathbf{j}\,\text{m s}^{-1}$ **b** $24\mathbf{i} - 7\mathbf{j}\,\text{km h}^{-1}$

  c $5\mathbf{i} + 2\mathbf{j}\,\text{m s}^{-1}$ **d** $-7\mathbf{i} + 4\mathbf{j}\,\text{cm s}^{-1}$

2 Find the distance moved by a particle which travels for:

  a 5 hours at velocity $8\mathbf{i} + 6\mathbf{j}\,\text{km h}^{-1}$

  b 10 seconds at velocity $5\mathbf{i} - \mathbf{j}\,\text{m s}^{-1}$

  c 45 minutes at velocity $6\mathbf{i} + 2\mathbf{j}\,\text{km h}^{-1}$

  d 2 minutes at velocity $-4\mathbf{i} - 7\mathbf{j}\,\text{cm s}^{-1}$.

3 Find the speed and the distance travelled by a particle moving with:

  a velocity $-3\mathbf{i} + 4\mathbf{j}\,\text{m s}^{-1}$ for 15 seconds

  b velocity $2\mathbf{i} + 5\mathbf{j}\,\text{m s}^{-1}$ for 3 seconds

  c velocity $5\mathbf{i} - 2\mathbf{j}\,\text{km h}^{-1}$ for 3 hours

  d velocity $12\mathbf{i} - 5\mathbf{j}\,\text{km h}^{-1}$ for 30 minutes.

6.6 You can solve problems involving velocity and time using vectors.

■ If a particle starts from the point with position vector $\mathbf{r_0}$ and moves with constant velocity $\mathbf{v}$, then its displacement from its initial position at time t is $\mathbf{v}t$ and its position vector $\mathbf{r}$ is given by

$$\mathbf{r} = \mathbf{r_0} + \mathbf{v}t$$

Example 12

A particle starts from the point with position vector $(3\mathbf{i} + 7\mathbf{j})$ m and moves with constant velocity $(2\mathbf{i} - \mathbf{j})\,\mathrm{m\,s^{-1}}$. Find the position vector of the particle 4 seconds later.

Displacement $= \mathbf{v}t = 4(2\mathbf{i} - \mathbf{j}) = 8\mathbf{i} - 4\mathbf{j}$ •———— Displacement $= \mathbf{v} \times t$

Position vector $\mathbf{r} = (3\mathbf{i} + 7\mathbf{j}) + (8\mathbf{i} - 4\mathbf{j})$ •———— $\mathbf{r} = \mathbf{r_0} + \mathbf{v}t$

$\qquad\qquad = (3 + 8)\mathbf{i} + (7 - 4)\mathbf{j}$ •

$\qquad\qquad = 11\mathbf{i} + 3\mathbf{j}$

Position vector after 4 seconds = position vector of starting point + displacement

Example 13

A particle moving at a constant velocity is at the point with position vector $(2\mathbf{i} - 4\mathbf{j})$ m at time $t = 0$. The particle is moving at a constant velocity. Five seconds later it is at the point with position vector $(12\mathbf{i} + 16\mathbf{j})$ m. Find the velocity of the particle.

Displacement $= (12\mathbf{i} + 16\mathbf{j}) - (2\mathbf{i} - 4\mathbf{j})$ •

$\qquad\qquad = (12 - 2)\mathbf{i} + (16 - (-4))\mathbf{j}$

$\qquad\qquad = 10\mathbf{i} + 20\mathbf{j}$

Travels $10\mathbf{i} + 20\mathbf{j}$ in 5 seconds, so

$\qquad \mathbf{v} = \frac{1}{5}(10\mathbf{i} + 20\mathbf{j}) = (2\mathbf{i} + 4\mathbf{j})\,\mathrm{m\,s^{-1}}$ •

You need to use the same formula $\mathbf{r} = \mathbf{r_0} + \mathbf{v}t$, but this time you are using it find $\mathbf{v}$.

Start by finding the displacement.

Travelling at constant velocity, so divide displacement by time taken to obtain $\mathbf{v}$.

Example 14

At time $t = 0$ a particle has position vector $4\mathbf{i} + 7\mathbf{j}$ and is moving with speed $15\,\mathrm{m\,s^{-1}}$ in the direction $3\mathbf{i} - 4\mathbf{j}$. Find its position vector after 2 seconds.

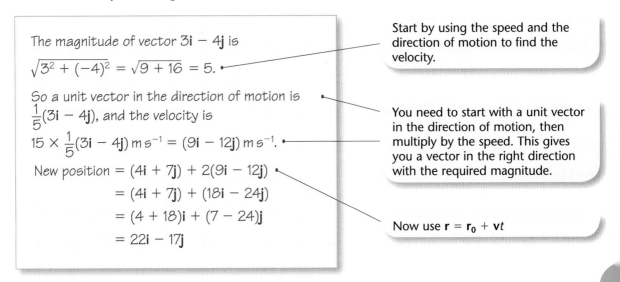

The magnitude of vector $3\mathbf{i} - 4\mathbf{j}$ is

$\sqrt{3^2 + (-4)^2} = \sqrt{9 + 16} = 5.$ •

So a unit vector in the direction of motion is
$\frac{1}{5}(3\mathbf{i} - 4\mathbf{j})$, and the velocity is

$15 \times \frac{1}{5}(3\mathbf{i} - 4\mathbf{j})\,\mathrm{m\,s^{-1}} = (9\mathbf{i} - 12\mathbf{j})\,\mathrm{m\,s^{-1}}.$ •

New position $= (4\mathbf{i} + 7\mathbf{j}) + 2(9\mathbf{i} - 12\mathbf{j})$ •

$\qquad\qquad = (4\mathbf{i} + 7\mathbf{j}) + (18\mathbf{i} - 24\mathbf{j})$

$\qquad\qquad = (4 + 18)\mathbf{i} + (7 - 24)\mathbf{j}$

$\qquad\qquad = 22\mathbf{i} - 17\mathbf{j}$

Start by using the speed and the direction of motion to find the velocity.

You need to start with a unit vector in the direction of motion, then multiply by the speed. This gives you a vector in the right direction with the required magnitude.

Now use $\mathbf{r} = \mathbf{r_0} + \mathbf{v}t$

■ The acceleration of a particle tells you how the velocity changes with time. Acceleration is a vector, usually denoted by **a**. If a particle with initial velocity **u** moves with constant acceleration **a** then its velocity, **v**, at time t is given by

$$\mathbf{v} = \mathbf{u} + \mathbf{a}t$$

Example 15

A particle P has velocity $(-3\mathbf{i} + \mathbf{j})\,\mathrm{m\,s^{-1}}$ at time $t = 0$. The particle moves with constant acceleration $\mathbf{a} = (2\mathbf{i} + 3\mathbf{j})\,\mathrm{m\,s^{-2}}$. Find the speed of the particle after 3 seconds.

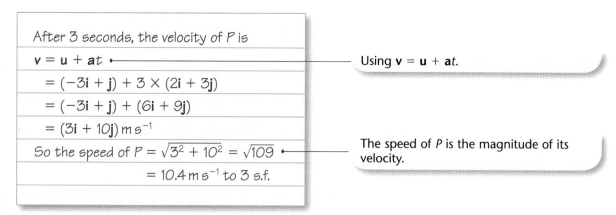

After 3 seconds, the velocity of P is

$\mathbf{v} = \mathbf{u} + \mathbf{a}t$ ———————————— Using $\mathbf{v} = \mathbf{u} + \mathbf{a}t$.

$\quad = (-3\mathbf{i} + \mathbf{j}) + 3 \times (2\mathbf{i} + 3\mathbf{j})$

$\quad = (-3\mathbf{i} + \mathbf{j}) + (6\mathbf{i} + 9\mathbf{j})$

$\quad = (3\mathbf{i} + 10\mathbf{j})\,\mathrm{m\,s^{-1}}$

So the speed of $P = \sqrt{3^2 + 10^2} = \sqrt{109}$ ——— The speed of P is the magnitude of its velocity.

$\quad\quad\quad = 10.4\,\mathrm{m\,s^{-1}}$ to 3 s.f.

■ A force applied to a particle has both a magnitude and a direction, so force is also a vector. The force causes the particle to accelerate:

$\mathbf{F} = m\mathbf{a}$, where m is the mass of the particle.

Example 16

A constant force, $\mathbf{F}\,\mathrm{N}$, acts on a particle of mass $2\,\mathrm{kg}$ for 10 seconds. The particle is initially at rest, and 10 seconds later it has velocity $(10\mathbf{i} - 24\mathbf{j})\,\mathrm{m\,s^{-1}}$. Find $\mathbf{F}$.

$(10\mathbf{i} - 24\mathbf{j}) = 10\mathbf{a},$ ———————— Use $\mathbf{v} = \mathbf{u} + \mathbf{a}t$ to find the acceleration of the particle. $\mathbf{u} = 0$ as the particle is initially at rest.

So $\mathbf{a} = (\mathbf{i} - 2.4\mathbf{j})\,\mathrm{m\,s^{-2}}.$

So $\mathbf{F} = 2 \times (\mathbf{i} - 2.4\mathbf{j}) = (2\mathbf{i} - 4.8\mathbf{j})\,\mathrm{N}$ ———— Using $\mathbf{F} = m\mathbf{a}$

Exercise 6F

1 A particle P is moving with constant velocity $\mathbf{v}\,\mathrm{m\,s^{-1}}$. Initially P is at the point with position vector $\mathbf{r}$. Find the position of P t seconds later if:

a $\mathbf{r} = 3\mathbf{j}$, $\mathbf{v} = 2\mathbf{i}$ and $t = 4$,　　　　　**b** $\mathbf{r} = 2\mathbf{i} - \mathbf{j}$, $\mathbf{v} = -2\mathbf{j}$ and $t = 3$,

c $\mathbf{r} = \mathbf{i} + 4\mathbf{j}$, $\mathbf{v} = -3\mathbf{i} + 2\mathbf{j}$ and $t = 6$,　　**d** $\mathbf{r} = -3\mathbf{i} + 2\mathbf{j}$, $\mathbf{v} = 2\mathbf{i} - 3\mathbf{j}$ and $t = 5$.

2 A particle P moves with constant velocity $\mathbf{v}$. Initially P is at the point with position vector $\mathbf{a}$. t seconds later P is at the point with position vector $\mathbf{b}$. Find $\mathbf{v}$ when:

a $\mathbf{a} = 2\mathbf{i} + 3\mathbf{j}$, $\mathbf{b} = 6\mathbf{i} + 13\mathbf{j}$, $t = 2$,

b $\mathbf{a} = 4\mathbf{i} + \mathbf{j}$, $\mathbf{b} = 9\mathbf{i} + 16\mathbf{j}$, $t = 5$,

c $\mathbf{a} = 3\mathbf{i} - 5\mathbf{j}$, $\mathbf{b} = 9\mathbf{i} + 7\mathbf{j}$, $t = 3$,

d $\mathbf{a} = -2\mathbf{i} + 7\mathbf{j}$, $\mathbf{b} = 4\mathbf{i} - 8\mathbf{j}$, $t = 3$,

e $\mathbf{a} = -4\mathbf{i} + \mathbf{j}$, $\mathbf{b} = -12\mathbf{i} - 19\mathbf{j}$, $t = 4$.

3 A particle moving with speed $v\,\mathrm{m\,s^{-1}}$ in direction $\mathbf{d}$ has velocity vector $\mathbf{v}$. Find $\mathbf{v}$ for these.

a $v = 10$, $\mathbf{d} = 3\mathbf{i} - 4\mathbf{j}$ **b** $v = 15$, $\mathbf{d} = -4\mathbf{i} + 3\mathbf{j}$

c $v = 7.5$, $\mathbf{d} = -6\mathbf{i} + 8\mathbf{j}$ **d** $v = 5\sqrt{2}$, $\mathbf{d} = \mathbf{i} + \mathbf{j}$

e $v = 2\sqrt{13}$, $\mathbf{d} = -2\mathbf{i} + 3\mathbf{j}$ **f** $v = \sqrt{68}$, $\mathbf{d} = 3\mathbf{i} - 5\mathbf{j}$

g $v = \sqrt{60}$, $\mathbf{d} = -4\mathbf{i} - 2\mathbf{j}$ **h** $v = 15$, $\mathbf{d} = -\mathbf{i} + 2\mathbf{j}$

4 A particle P starts at the point with position vector $\mathbf{r_0}$. P moves with constant velocity $\mathbf{v}\,\mathrm{m\,s^{-1}}$. After t seconds, P is at the point with position vector $\mathbf{r}$.

a Find $\mathbf{r}$ if $\mathbf{r_0} = 2\mathbf{i}$, $\mathbf{v} = \mathbf{i} + 3\mathbf{j}$, and $t = 4$.

b Find $\mathbf{r}$ if $\mathbf{r_0} = 3\mathbf{i} - \mathbf{j}$, $\mathbf{v} = -2\mathbf{i} + \mathbf{j}$, and $t = 5$.

c Find $\mathbf{r_0}$ if $\mathbf{r} = 4\mathbf{i} + 3\mathbf{j}$, $\mathbf{v} = 2\mathbf{i} - \mathbf{j}$, and $\mathbf{t} = 3$.

d Find $\mathbf{r_0}$ if $\mathbf{r} = -2\mathbf{i} + 5\mathbf{j}$, $\mathbf{v} = -2\mathbf{i} + 3\mathbf{j}$, and $t = 6$.

e Find $\mathbf{v}$ if $\mathbf{r_0} = 2\mathbf{i} + 2\mathbf{j}$, $\mathbf{r} = 8\mathbf{i} - 7\mathbf{j}$, and $t = 3$.

f Find the speed of P if $\mathbf{r_0} = 10\mathbf{i} - 5\mathbf{j}$, $\mathbf{r} = -2\mathbf{i} + 9\mathbf{j}$, and $t = 4$.

g Find t if $\mathbf{r_0} = 4\mathbf{i} + \mathbf{j}$, $\mathbf{r} = 12\mathbf{i} - 11\mathbf{j}$, and $\mathbf{v} = 2\mathbf{i} - 3\mathbf{j}$.

h Find t if $\mathbf{r_0} = -2\mathbf{i} + 3\mathbf{j}$, $\mathbf{r} = 6\mathbf{i} - 3\mathbf{j}$, and the speed of P is $4\,\mathrm{m\,s^{-1}}$.

5 The initial velocity of a particle P moving with uniform acceleration $\mathbf{a}\,\mathrm{m\,s^{-2}}$ is $\mathbf{u}\,\mathrm{m\,s^{-1}}$. Find the velocity and the speed of P after t seconds in these cases.

a $\mathbf{u} = 5\mathbf{i}$, $\mathbf{a} = 3\mathbf{j}$, and $t = 4$ **b** $\mathbf{u} = 3\mathbf{i} - 2\mathbf{j}$, $\mathbf{a} = \mathbf{i} - \mathbf{j}$, and $t = 3$

c $\mathbf{a} = 2\mathbf{i} - 3\mathbf{j}$, $\mathbf{u} = -2\mathbf{i} + \mathbf{j}$, and $t = 2$ **d** $t = 6$, $\mathbf{u} = 3\mathbf{i} - 2\mathbf{j}$, and $\mathbf{a} = -\mathbf{i}$

e $\mathbf{a} = 2\mathbf{i} + \mathbf{j}$, $t = 5$, and $\mathbf{u} = -3\mathbf{i} + 4\mathbf{j}$

6 A constant force $\mathbf{F}\,\mathrm{N}$ acts on a particle of mass $4\,\mathrm{kg}$ for 5 seconds. The particle was initially at rest, and after 5 seconds it has velocity $6\mathbf{i} - 8\mathbf{j}\,\mathrm{m\,s^{-1}}$. Find $\mathbf{F}$.

7 A force $2\mathbf{i} - \mathbf{j}\,\mathrm{N}$ acts on a particle of mass $2\,\mathrm{kg}$. If the initial velocity of the particle is $\mathbf{i} + 3\mathbf{j}\,\mathrm{m\,s^{-1}}$, find how far it moves in the first 3 seconds.

8 At time $t = 0$, the particle P is at the point with position vector $4\mathbf{i}$, and moving with constant velocity $\mathbf{i} + \mathbf{j}\,\mathrm{m\,s^{-1}}$. A second particle Q is at the point with position vector $-3\mathbf{j}$ and moving with velocity $\mathbf{v}\,\mathrm{m\,s^{-1}}$. After 8 seconds, the paths of P and Q meet. Find the speed of Q.

9 At 2 pm the coastguard spots a rowing dinghy 500 m due south of his observation point. The dinghy has constant velocity $(2\mathbf{i} + 3\mathbf{j})\,\mathrm{m\,s^{-1}}$.

> In questions 9 and 10 the unit vectors **i** and **j** are due east and due north respectively.

a Find, in terms of t, the position vector of the dinghy t seconds after 2 pm.

b Find the distance of the dinghy from the observation point at 2.05 pm.

10 At noon a ferry F is 400 m due north of an observation point O moving with constant velocity $(7\mathbf{i} + 7\mathbf{j})\,\mathrm{m\,s^{-1}}$, and a speedboat S is 500 m due east of O, moving with constant velocity $(-3\mathbf{i} + 15\mathbf{j})\,\mathrm{m\,s^{-1}}$.

a Write down the position vectors of F and S at time t seconds after noon.

b Show that F and S will collide, and find the position vector of the point of collision.

11 At 8 am two ships A and B are at $\mathbf{r}_A = (\mathbf{i} + 3\mathbf{j})$ km and $\mathbf{r}_B = (5\mathbf{i} - 2\mathbf{j})$ km from a fixed point P. Their velocities are $\mathbf{v}_A = (2\mathbf{i} - \mathbf{j})\,\mathrm{km\,h^{-1}}$ and $\mathbf{v}_B = (-\mathbf{i} + 4\mathbf{j})\,\mathrm{km\,h^{-1}}$ respectively.

a Write down the position vectors of A and B t hours later.

b Show that t hours after 8 am the position vector of B relative to A is given by $((4 - 3t)\mathbf{i} + (-5 + 5t)\mathbf{j})$ km.

c Show that the two ships do not collide.

d Find the distance between A and B at 10 am.

12 A particle A starts at the point with position vector $12\mathbf{i} + 12\mathbf{j}$. The initial velocity of A is $(-\mathbf{i} + \mathbf{j})\,\mathrm{m\,s^{-1}}$, and it has constant acceleration $(2\mathbf{i} - 4\mathbf{j})\,\mathrm{m\,s^{-2}}$. Another particle, B, has initial velocity $\mathbf{i}\,\mathrm{m\,s^{-1}}$ and constant acceleration $2\mathbf{j}\,\mathrm{m\,s^{-2}}$. After 3 seconds the two particles collide. Find

a the speeds of the two particles when they collide.

b the position vector of the point where the two particles collide,

c the position vector of B's starting point.

6.7 You can use vectors to solve problems about forces.

■ **If a particle is resting in equilibrium then the resultant of all the forces acting on it is zero. This means the sum of the vectors of the forces is the zero vector.**

Example **17**

The forces $2\mathbf{i} + 3\mathbf{j}$, $4\mathbf{i} - \mathbf{j}$, $-3\mathbf{i} + 2\mathbf{j}$ and $a\mathbf{i} + b\mathbf{j}$ act on a particle which is in equilibrium. Find the values of a and b.

$(2\mathbf{i} + 3\mathbf{j}) + (4\mathbf{i} - \mathbf{j}) + (-3\mathbf{i} + 2\mathbf{j}) + (a\mathbf{i} + b\mathbf{j}) = 0$

> If the particle is in equilibrium then the resultant force will be zero.

$(2 + 4 - 3 + a)\mathbf{i} + (3 - 1 + 2 + b)\mathbf{j} = 0$

$\Rightarrow \quad 3 + a = 0 \text{ and } 4 + b = 0$

$\Rightarrow \qquad a = -3 \quad \text{and} \quad b = -4$

Example 18

Forces $\mathbf{F}_1 = (2\mathbf{i} + 4\mathbf{j})\,\text{N}$, $\mathbf{F}_2 = (-5\mathbf{i} + 4\mathbf{j})\,\text{N}$, and $\mathbf{F}_3 = (6\mathbf{i} - 5\mathbf{j})\,\text{N}$ act on a particle of mass 3 kg. Find the acceleration of the particle.

Resultant force

$$= \mathbf{F}_1 + \mathbf{F}_2 + \mathbf{F}_3$$

Add the vectors to find the resultant force.

$$= (2\mathbf{i} + 4\mathbf{j}) + (-5\mathbf{i} + 4\mathbf{j}) + (6\mathbf{i} - 5\mathbf{j})$$

$$= 3\mathbf{i} + 3\mathbf{j}$$

$$3\mathbf{i} + 3\mathbf{j} = 3\mathbf{a} \quad \Rightarrow \quad \mathbf{a} = (\mathbf{i} + \mathbf{j})\,)\,\text{ms}^{-2}$$

Use $F = ma$.

Example 19

A particle P, of weight 7 N is suspended in equilibrium by two light inextensible strings attached to P and to points A and B, as shown in the diagram. The line AB is horizontal. Find the tensions in the two strings.

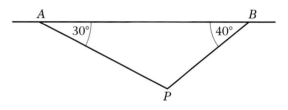

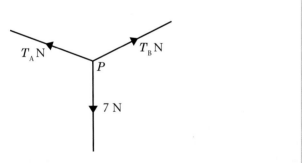

Draw a diagram showing the forces acting on P.

Triangle of forces:

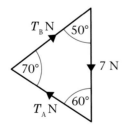

Since the particle is hanging in equilibrium, there is no resultant force. This means that if we add the three force vectors together the resultant is zero. This means that the vectors form a triangle.

$$\Rightarrow \quad \frac{7}{\sin 70} = \frac{T_A}{\sin 50} = \frac{T_B}{\sin 60}$$

Use the Sine rule to find the unknown forces.

$$\Rightarrow \quad T_A = \frac{7 \times \sin 50}{\sin 70} = 5.71\,\text{N (3 s.f.)}$$

and $$T_B = \frac{7 \times \sin 60}{\sin 70} = 6.45\,\text{N (3 s.f.)}$$

Exercise 6G

1 A particle is in equilibrium at O under the action of three forces $\mathbf{F}_1$, $\mathbf{F}_2$ and $\mathbf{F}_3$. Find $\mathbf{F}_3$ in these cases.

a $\mathbf{F}_1 = (2\mathbf{i} + 7\mathbf{j})$ and $\mathbf{F}_2 = (-3\mathbf{i} + \mathbf{j})$ **b** $\mathbf{F}_1 = (3\mathbf{i} - 4\mathbf{j})$ and $\mathbf{F}_2 = (2\mathbf{i} + 3\mathbf{j})$

c $\mathbf{F}_1 = (-4\mathbf{i} - 2\mathbf{j})$ and $\mathbf{F}_2 = (2\mathbf{i} - 3\mathbf{j})$ **d** $\mathbf{F}_1 = (-\mathbf{i} - 3\mathbf{j})$ and $\mathbf{F}_2 = (4\mathbf{i} + \mathbf{j})$

2 For each part of Question 1 find the magnitude of $\mathbf{F}_3$ and the angle it makes with the positive x-axis.

3 Forces $\mathbf{P}$N, $\mathbf{Q}$N and $\mathbf{R}$N act on a particle of m kg. Find the resultant force on the particle and the acceleration produced when

a $\mathbf{P} = 3\mathbf{i} + \mathbf{j}$, $\mathbf{Q} = 2\mathbf{i} - 3\mathbf{j}$, $\mathbf{R} = \mathbf{i} + 2\mathbf{j}$ and $m = 2$,

b $\mathbf{P} = 4\mathbf{i} - 3\mathbf{j}$, $\mathbf{Q} = -3\mathbf{i} + 2\mathbf{j}$, $\mathbf{R} = 2\mathbf{i} - \mathbf{j}$ and $m = 3$,

c $\mathbf{P} = -3\mathbf{i} + 2\mathbf{j}$, $\mathbf{Q} = 2\mathbf{i} - 5\mathbf{j}$, $\mathbf{R} = 4\mathbf{i} + \mathbf{j}$ and $m = 4$,

d $\mathbf{P} = 2\mathbf{i} + \mathbf{j}$, $\mathbf{Q} = -6\mathbf{i} - 4\mathbf{j}$, $\mathbf{R} = 5\mathbf{i} - 3\mathbf{j}$ and $m = 2$.

4

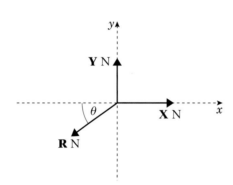

A particle is in equilibrium at O under the action of forces $\mathbf{X}$, $\mathbf{Y}$ and $\mathbf{R}$, as shown in the diagram. Use a triangle of forces to find the magnitude of $\mathbf{R}$ and the value of θ when:

a $|\mathbf{X}| = 3$ N, $|\mathbf{Y}| = 5$ N,   **b** $|\mathbf{X}| = 6$ N, $|\mathbf{Y}| = 2$ N,   **c** $|\mathbf{X}| = 5$ N, $|\mathbf{Y}| = 4$ N.

5

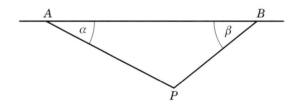

The diagram shows two strings attached to a particle P, of weight W N, and to two fixed points A and B. The line AB is horizontal and P is hanging in equilibrium with $\angle BAP = \alpha$ and $\angle ABP = \beta$. The magnitude of the tension in AP is T_A N, and the magnitude of the tension in BP is T_B N.

Use a triangle of vectors to find:

a T_A and T_B if $W = 5$, $\alpha = 40°$ and $\beta = 50°$,   **b** T_A and T_B if $W = 6$, $\alpha = 30°$ and $\beta = 45°$,

c T_A and W if $T_B = 5$, $\alpha = 40°$ and $\beta = 50°$,   **d** W and T_B if $T_A = 5$, $\alpha = 30°$ and $\beta = 70°$,

e T_A and α if $W = 7$, $T_B = 6$ and $\beta = 50°$.

Mixed exercise 6H

1 Three forces $\mathbf{F}_1$, $\mathbf{F}_2$ and $\mathbf{F}_3$ act on a particle. $\mathbf{F}_1 = (-3\mathbf{i} + 7\mathbf{j})\,\text{N}$, $\mathbf{F}_2 = (\mathbf{i} - \mathbf{j})\,\text{N}$ and $\mathbf{F}_3 = (p\mathbf{i} + q\mathbf{j})\,\text{N}$.

  a Given that this particle is in equilibrium, determine the value of p and the value of q.

  The resultant of the forces $\mathbf{F}_1$ and $\mathbf{F}_2$ is $\mathbf{R}$.

  b Calculate, in N, the magnitude of $\mathbf{R}$.

  c Calculate, to the nearest degree, the angle between the line of action of $\mathbf{R}$ and the vector $\mathbf{j}$.

2 A coastguard station O monitors the movements of ships in a channel. At noon, the station's radar records two ships moving with constant speed. Ship A is at the point with position vector $(-3\mathbf{i} + 10\mathbf{j})\,\text{km}$ relative to O and has velocity $(2\mathbf{i} + 2\mathbf{j})\,\text{km h}^{-1}$. Ship B is at the point with position vector $(6\mathbf{i} + \mathbf{j})\,\text{km}$ and has velocity $(-\mathbf{i} + 5\mathbf{j})\,\text{km h}^{-1}$.

> In this question, the horizontal unit vectors $\mathbf{i}$ and $\mathbf{j}$ are directed due east and north respectively.

  a Show that if the two ships maintain these velocities they will collide.

  The coastguard radios ship A and orders it to reduce its speed to move with velocity $(\mathbf{i} + \mathbf{j})\,\text{km h}^{-1}$. Given that A obeys this order and maintains this new constant velocity.

  b find an expression for the vector $\overrightarrow{AB}$ at time t hours after noon,

  c find, to three significant figures, the distance between A and B at 1500 hours,

  d find the time at which B will be due north of A.

3 Two ships P and Q are moving along straight lines with constant velocities. Initially P is at a point O and the position vector of Q relative to O is $(12\mathbf{i} + 6\mathbf{j})\,\text{km}$, where $\mathbf{i}$ and $\mathbf{j}$ are unit vectors directed due east and due north respectively. Ship P is moving with velocity $6\mathbf{i}\,\text{km h}^{-1}$ and ship Q is moving with velocity $(-3\mathbf{i} + 6\mathbf{j})\,\text{km h}^{-1}$. At time t hours the position vectors of P and Q relative to O are $\mathbf{p}\,\text{km}$ and $\mathbf{q}\,\text{km}$ respectively.

  a Find $\mathbf{p}$ and $\mathbf{q}$ in terms of t.

  b Calculate the distance of Q from P when $t = 4$.

  c Calculate the value of t when Q is due north of P.

4 A particle P moves with constant acceleration $(-3\mathbf{i} + \mathbf{j})\,\text{m s}^{-2}$. At time t seconds, its velocity is $\mathbf{v}\,\text{m s}^{-1}$. When $t = 0$, $\mathbf{v} = 5\mathbf{i} - 3\mathbf{j}$.

  a Find the value of t when P is moving parallel to the vector $\mathbf{i}$.

  b Find the speed of P when $t = 5$.

  c Find the angle between the vector $\mathbf{i}$ and the direction of motion of P when $t = 5$.

5 A particle P of mass $5\,\text{kg}$ is moving under the action of a constant force $\mathbf{F}$ newtons. At $t = 0$, P has velocity $(5\mathbf{i} - 3\mathbf{j})\,\text{m s}^{-1}$. At $t = 4\,\text{s}$, the velocity of P is $(-11\mathbf{i} + 5\mathbf{j})\,\text{m s}^{-1}$. Find

  a the acceleration of P in terms of $\mathbf{i}$ and $\mathbf{j}$,

  b the magnitude of $\mathbf{F}$.

  At $t = 6\,\text{s}$, P is at the point A with position vector $(28\mathbf{i} + 6\mathbf{j})\,\text{m}$ relative to a fixed origin O. At this instant the force $\mathbf{F}$ newtons is removed and P then moves with constant velocity. Two seconds after the force has been removed, P is at the point B.

  c Calculate the distance of B from O.

6 Two boats A and B are moving with constant velocities. Boat A moves with velocity $6\mathbf{i}$ km h^{-1}. Boat B moves with velocity $(3\mathbf{i} + 5\mathbf{j})$ km h^{-1}.

> In this question the vectors $\mathbf{i}$ and $\mathbf{j}$ are horizontal unit vectors in the directions due east and due north respectively.

a Find the bearing on which B is moving.

At noon, A is at point O and B is 10 km due south of O. At time t hours after noon, the position vectors of A and B relative to O are $\mathbf{a}$ km and $\mathbf{b}$ km respectively.

b Find expressions for $\mathbf{a}$ and $\mathbf{b}$ in terms of t, giving your answer in the form $p\mathbf{i} + q\mathbf{j}$.

c Find the time when A is due east of B.

At time t hours after noon, the distance between A and B is d km. By finding an expression for $\overrightarrow{AB}$,

d show that $d^2 = 34t^2 - 100t + 100$.

At noon, the boats are 10 km apart.

e Find the time after noon at which the boats are again 10 km apart.

7 A small boat S, drifting in the sea, is modelled as a particle moving in a straight line at constant speed. When first sighted at 0900, S is at a point with position vector $(-2\mathbf{i} - 4\mathbf{j})$ km relative to a fixed origin O, where $\mathbf{i}$ and $\mathbf{j}$ are unit vectors due east and due north respectively. At 0940, S is at the point with position vector $(4\mathbf{i} - 6\mathbf{j})$ km. At time t hours after 0900, S is at the point with position vector $\mathbf{s}$ km.

a Calculate the bearing on which S is drifting.

b Find an expression for $\mathbf{s}$ in terms of t.

At 1100 a motor boat M leaves O and travels with constant velocity $(p\mathbf{i} + q\mathbf{j})$ km h^{-1}.

c Given that M intercepts S at 1130, calculate the value of p and the value of q.

8 A particle P moves in a horizontal plane. The acceleration of P is $(-2\mathbf{i} + 3\mathbf{j})$ m s^{-2}. At time $t = 0$, the velocity of P is $(3\mathbf{i} - 2\mathbf{j})$ m s^{-1}.

a Find, to the nearest degree, the angle between the vector $\mathbf{j}$ and the direction of motion of P when $t = 0$.

At time t seconds, the velocity of P is $\mathbf{v}$ m s^{-1}. Find

b an expression for $\mathbf{v}$ in terms of t, in the form $a\mathbf{i} + b\mathbf{j}$,

c the speed of P when $t = 4$,

d the time when P is moving parallel to $\mathbf{i} + \mathbf{j}$.

9 At time $t = 0$ a football player kicks a ball from the point A with position vector $(3\mathbf{i} + 2\mathbf{j})$ m on a horizontal football field. The motion of the ball is modelled as that of a particle moving horizontally with constant velocity $(4\mathbf{i} + 9\mathbf{j})$ m s^{-1}. Find

> In this question, the unit vectors $\mathbf{i}$ and $\mathbf{j}$ are horizontal vectors due east and north respectively.

a the speed of the ball,

b the position vector of the ball after t seconds.

The point B on the field has position vector $(29\mathbf{i} + 12\mathbf{j})$ m.

c Find the time when the ball is due north of B.

At time $t = 0$, another player starts running due north from B and moves with constant speed v m s^{-1}.

d Given that he intercepts the ball, find the value of v.

10 Two ships P and Q are travelling at night with constant velocities. At midnight, P is at the point with position vector $(10\mathbf{i} + 15\mathbf{j})$ km relative to a fixed origin O. At the same time, Q is at the point with position vector $(-16\mathbf{i} + 26\mathbf{j})$ km. Three hours later, P is at the point with position vector $(25\mathbf{i} + 24\mathbf{j})$ km. The ship Q travels with velocity $12\mathbf{i}$ km h^{-1}. At time t hours after midnight, the position vectors of P and Q are $\mathbf{p}$ km and $\mathbf{q}$ km respectively. Find

a the velocity of P in terms of $\mathbf{i}$ and $\mathbf{j}$,

b expressions for $\mathbf{p}$ and $\mathbf{q}$ in terms of t, $\mathbf{i}$ and $\mathbf{j}$.

At time t hours after midnight, the distance between P and Q is d km.

c By finding an expression for $\overrightarrow{PQ}$, show that
$$d^2 = 58t^2 - 430t + 797$$

Weather conditions are such that an observer on P can only see the lights on Q when the distance between P and Q is 13 km or less.

d Given that when $t = 2$ the lights on Q move into sight of the observer, find the time, to the nearest minute, at which the lights on Q move out of sight of the observer.

$\times (2u, a)$

$B'(2a, 12)$

Summary of key points

1 A vector is a quantity which has both magnitude and direction.

2 A vector can be represented as a directed line segment.

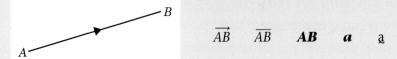

$\overrightarrow{AB}$ $\overline{AB}$ **AB**   *a*   a̰

3 Two vectors are equal if and only if they have the same magnitude and the same direction.

4 Two vectors are parallel if and only if they have the same direction.

5 You can add vectors using the triangle of law of addition.

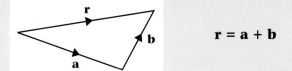

r = a + b

6 The unit vectors along the Cartesian axes are usually denoted by **i** and **j** respectively. You can write any two-dimensional vector in the form $a\mathbf{i} + b\mathbf{j}$.

7 When vectors are written in terms of the unit vectors **i** and **j** you can add them together by adding the terms in **i** and the terms in **j** separately. You subtract vectors in a similar way.

8 When a vector is given in terms of the unit vectors **i**, and **j** you can find its magnitude using Pythagoras' Theorem. The magnitude of a vector **a** is written $|\boldsymbol{a}|$.

9 The velocity of a particle is a vector in the direction of motion. The magnitude of the velocity is the speed of the particle. The velocity is usually denoted by **v**.

10 If a particle starts from the point with position vector $\mathbf{r}_0$ and moves with constant velocity **v**, then its displacement from its initial position at time t is **v**t and its position vector **r** is given by

$$\mathbf{r} = \mathbf{r}_0 + \mathbf{v}t.$$

11 The acceleration of a particle tells you how the velocity changes with time. Acceleration is a vector, usually denoted by **a**. If a particle with initial velocity **u** moves with constant acceleration **a** then its velocity, **v**, at time t is given by

$$\mathbf{v} = \mathbf{u} + \mathbf{a}t.$$

12 A force applied to a particle has both a magnitude and a direction, so force is also a vector. The force causes the particle to accelerate:

$$\mathbf{F} = m\mathbf{a}, \text{ where } m \text{ is the mass of the particle}$$

13 If a particle is resting in equilibrium then the resultant of all the forces acting on it is zero. This means that the sum of the vectors of the forces is the zero vector.

1

A particle of weight 24 N is held in equilibrium by two light inextensible strings. One string is horizontal. The other string is inclined at an angle of 30° to the horizontal, as shown. The tension in the horizontal string is Q newtons and the tension in the other string is P newtons. Find

a the value of P,

b the value of Q. **E**

2

A particle P, of mass 2 kg, is attached to one end of a light string, the other end of which is attached to a fixed point O. The particle is held in equilibrium, with OP horizontal, by a force of magnitude 30 N applied at an angle α to the horizontal, as shown.

a Find, to the nearest degree, the value of α.

b Find, in N to 3 significant figures, the magnitude of the tension in the string. **E**

3

A body of mass 5 kg is held in equilibrium under gravity by two inextensible light ropes. One rope is horizontal, the other is at an angle α to the horizontal, as shown. The tension in the rope inclined at α to the horizontal is 72 N. Find

a the angle α, giving your answer to the nearest degree,

b the tension T in the horizontal rope, giving your answer to the nearest N. **E**

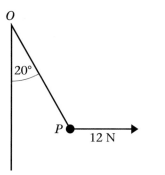

A particle P is attached to one end of a light inextensible string. The other end of the string is attached to a fixed point O. A horizontal force of magnitude 12 N is applied to P. The particle P is in equilibrium with the string taut and OP making an angle of 20° with the downward vertical, as shown. Find

a the tension in the string,

b the weight of P. **E**

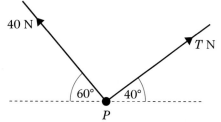

A particle P is held in equilibrium under gravity by two light, inextensible strings. One string is inclined at an angle of 60° to the horizontal and has a tension of 40 N. The other string is inclined at an angle of 40° to the horizontal and has a tension of T newtons, as shown. Find, to three significant figures,

a the value of T,

b the weight of P. **E**

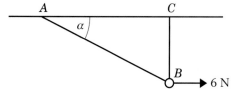

A smooth bead B is threaded on a light inextensible string. The ends of the string are attached to two fixed points A and C

on the same horizontal level. The bead is held in equilibrium by a horizontal force of magnitude 6 N acting parallel to AC. The bead B is vertically below C and $\angle BAC = \alpha$, as shown in the diagram. Given that $\tan \alpha = \frac{3}{4}$, find

a the tension in the string,

b the weight of the bead. **E**

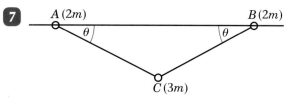

Two small rings, A and B, each of mass $2m$, are threaded on a rough horizontal pole. The coefficient of friction between each ring and the pole is μ. The rings are attached to the ends of a light inextensible string. A smooth ring C, of mass $3m$, is threaded on the string and hangs in equilibrium below the pole. The rings A and B are in limiting equilibrium on the pole, with $\angle BAC = \angle ABC = \theta$, where $\tan \theta = \frac{3}{4}$, as shown in the diagram.

a Show that the tension in the string is $\frac{5}{2}mg$.

b Find the value of μ. **E**

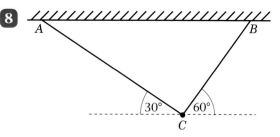

A particle of weight W newtons is attached at C to the ends of two light inextensible strings AC and BC. The other ends of the strings are attached to two fixed points A and B on a horizontal ceiling. The particle hangs in equilibrium with AC and BC inclined to the horizontal at 30° and 60° respectively, as shown. Given the tension in AC is 50 N, calculate

a the tension in BC, to three significant figures,

b the value of W.

 **9**

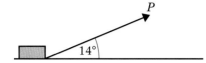

A small box of mass 20 kg rests on a rough horizontal floor. The coefficient of friction between the box and the floor is 0.25. A light inextensible rope is tied to the box and pulled with a force of magnitude P newtons at 14° to the horizontal as shown in the diagram. Given that the box is on the point of sliding, find the value of P, giving your answer to 1 decimal place. **E**

10

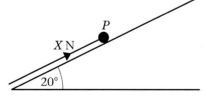

A smooth plane is inclined at an angle 10° to the horizontal. A particle P of mass 2 kg is held in equilibrium on the plane by a horizontal force of magnitude F newtons, as shown.

Find, to three significant figures,

a the normal reaction exerted by the plane on P.

b the value of F. **E**

11

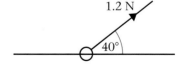

A particle P of mass 2.5 kg rests in equilibrium on a rough plane under the action of a force of magnitude X newtons acting up a line of greatest slope of the plane, as shown in the diagram. The plane

is inclined at 20° to the horizontal. The coefficient of friction between P and the plane is 0.4. The particle is in limiting equilibrium and is on the point of moving up the plane. Calculate

a the normal reaction of the plane on P,

b the value of X.

The force of magnitude X newtons is now removed.

c Show that P remains in equilibrium on the plane. **E**

12

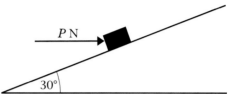

A parcel of weight 10 N lies on a rough plane inclined at an angle of 30° to the horizontal. A horizontal force of magnitude P newtons acts on the parcel, as shown. The parcel is in equilibrium and on the point of slipping up the plane. The normal reaction of the plane on the parcel is 18 N. The coefficient of friction between the parcel and the plane is μ. Find

a the value of P,

b the value of μ.

The horizontal force is removed.

c Determine whether or not the parcel moves. **E**

13

A small ring of mass 0.25 kg is threaded on a fixed rough horizontal rod. The ring is pulled upwards by a light string which makes an angle 40° with the horizontal, as shown. The string and the rod are in the same vertical plane. The tension in the string is 1.2 N and the coefficient of friction between the ring and the rod is μ. Given that the ring is in limiting equilibrium, find

a the normal reaction between the ring and the rod,

b the value of μ.

14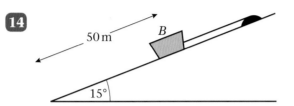

The diagram shows a boat B of mass 400 kg held at rest on a slipway by a rope. The boat is modelled as a particle and the slipway as a rough plane inclined at 15° to the horizontal. The coefficient of friction between B and the slipway is 0.2. The rope is modelled as a light, inextensible string, parallel to a line of greatest slope of the plane. The boat is in equilibrium and on the point of sliding down the slipway.

a Calculate the tension in the rope.

The boat is 50 m from the bottom of the slipway. The rope is detached from the boat and the boat slides down the slipway.

b Calculate the time taken for the boat to slide to the bottom of the slipway.

15

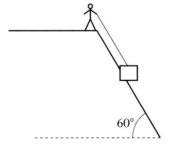

A heavy package is held in equilibrium on a slope by a rope. The package is attached to one end of the rope, the other end being held by a man standing at the top of the slope. The package is modelled as a particle of mass 20 kg. The slope is modelled as a rough plane inclined at 60° to the horizontal and the rope as a light inextensible string. The string is assumed to be parallel to a line of greatest slope of the plane, as shown in the diagram. At the contact between the package and the slope, the coefficient of friction is 0.4.

a Find the minimum tension in the rope for the package to stay in equilibrium on the slope.

The man now pulls the package up the slope. Given that the package moves at constant speed,

b find the tension in the rope.

c State how you have used, in your answer to part **b**, the fact that the package moves
  i up the slope,
  ii at constant speed.

16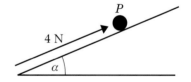

A particle P of mass 0.5 kg is on a rough plane inclined at an angle α to the horizontal, where $\tan \alpha = \frac{3}{4}$. The particle is held at rest on the plane by the action of a force of magnitude 4 N acting up the plane in a direction parallel to a line of greatest slope of the plane, as shown. The particle is on the point of slipping up the plane.

a Find the coefficient of friction between P and the plane.

The force of magnitude 4 N is removed.

b Find the acceleration of P down the plane.

A box of mass 30 kg is being pulled along rough horizontal ground at a constant speed using a rope. The rope makes an angle of 20° with the ground, as shown. The coefficient of friction between the box and the ground is 0.4. The box is modelled as a particle and the rope as a light, inextensible string. The tension in the rope is P newtons.

a Find the value of P.

The tension in the rope is now increased to 150 N.

b Find the acceleration of the box. **E**

18

A uniform rod AB has length 8 m and mass 12 kg. A particle of mass 8 kg is attached to the rod at B. The rod is supported at a point C and is in equilibrium in a horizontal position, as shown.

Find the length of AC. **E**

19

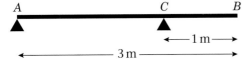

A uniform beam AB has mass 12 kg and length 3 m. The beam rests in equilibrium in a horizontal position, resting on two smooth supports. One support is at end A, the other at a point C on the beam, where $BC = 1$ m, as shown in the diagram. The beam is modelled as a uniform rod.

a Find the reaction on the beam at C.

A woman of mass 48 kg stands on the beam at the point D. The beam remains in equilibrium. The reactions on the beam at A and C are now equal.

b Find the distance AD. **E**

20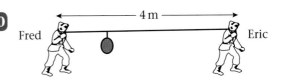

Two men, Eric and Fred, set out to carry a water container across a desert, using a long uniform pole. The length of the pole is 4 m and its mass is 5 kg. The ends of the pole rest in equilibrium on the shoulders of the two men, with the pole horizontal. The water container has mass 16 kg and is suspended from the pole by means of a light rope, which is short enough to prevent the container reaching the ground, as shown. Eric has a sprained ankle, so Fred fixes the rope in such a way that the vertical force on his shoulder is twice as great as the vertical force on Eric's shoulder.

a Find the vertical force on Eric's shoulder.

b Find the distance from the centre of the pole to the point at which the rope is fixed. **E**

21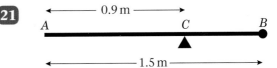

A uniform rod AB has length 1.5 m and mass 8 kg. A particle of mass m kg is attached to the rod at B. The rod is supported at the point C, where $AC = 0.9$ m, and the system is in equilibrium with AB horizontal, as shown.

a Show that $m = 2$.

A particle of mass 5 kg is now attached to the rod at A and the support is moved from C to a point D of the rod. The system, including both particles, is again in equilibrium with AB horizontal.

b Find the distance AD. **E**

22

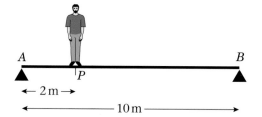

A uniform steel girder *AB*, of mass 150 kg and length 10 m, rests horizontally on two supports at *A* and *B*. A man of mass 90 kg stands on the girder at the point *P*, where *AP* = 2 m, as shown. By modelling the girder as a uniform rod and the man as a particle,

a find the magnitude of the reaction at *B*.

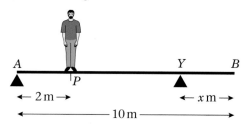

The support *B* is moved to a point *Y* on the girder, where *BY* = *x* metres, as shown. The man remains on the girder at *P*. The magnitudes of the reactions at the two supports are now equal.

Find

b the magnitude of the reaction at each support,

c the value of *x*.

23 A footbridge across a stream consists of a uniform horizontal plank *AB* of length 5 m and mass 140 kg, supported at the ends *A* and *B*. A man of mass 100 kg is standing at a point *C* on the footbridge. Given that the magnitude of the force exerted by the support at *A* is twice the magnitude of the force exerted by the support at *B*, calculate

a the magnitude, in N, of the force exerted by the support at *B*,

b the distance *AC*.

24 A non-uniform thin straight rod *AB* has length 3*d* and mass 5*m*. It is in equilibrium resting horizontally on supports at the points *X* and *Y*, where *AX* = *XY* = *YB* = *d*. A particle of mass 2*m* is attached to the rod at *B*. Given that the rod is on the point of tilting about *Y*, find the distance of the centre of mass of the rod from *A*.

25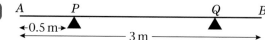

A uniform rod *AB* has weight 70 N and length 3 m. It rests in a horizontal position on two smooth supports placed at *P* and *Q*, where *AP* = 0.5 m as shown in the diagram. The reaction on the rod at *P* has magnitude 20 N. Find

a the magnitude of the reaction on the rod at *Q*,

b the distance *AQ*.

26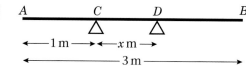

A uniform plank *AB* has weight 120 N and length 3 m. The plank rests horizontally in equilibrium on two smooth supports *C* and *D*, where *AC* = 1 m and *CD* = *x* m, as shown. The reaction of the support on the plank at *D* has magnitude 80 N. Modelling the plank as a rod.

a show that *x* = 0.75.

A rock is now placed at *B* and the plank is on the point of tilting about *D*. Modelling the rock as a particle, find

b the weight of the rock,

c the magnitude of the reaction of the support on the plank at *D*.

d State how you have used the model of the rock as a particle.

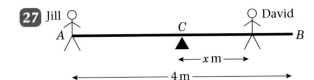

27 A seesaw in a playground consists of a beam AB of length 4 m which is supported by a smooth pivot at its centre C. Jill has mass 25 kg and sits on the end A. David has mass 40 kg and sits at a distance x metres from C, as shown. The beam is initially modelled as a uniform rod. Using this model,

a find the value of x for which the seesaw can rest in equilibrium in a horizontal position.

b State what is implied by the modelling assumptions that the beam is uniform.

David realises that the beam is not uniform as he finds he must sit at a distance 1.4 m from C for the seesaw to rest horizontally in equilibrium. The beam is now modelled as a non-uniform rod of mass 15 kg. Using this model,

c find the distance of the centre of mass of the beam from C. **E**

28

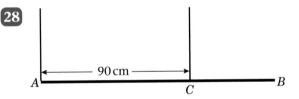

A steel girder AB has weight 210 N. It is held in equilibrium in a horizontal position by two vertical cables. One cable is attached to the end A. The other cable is attached to the point C on the girder, where $AC = 90$ cm, as shown. The girder is modelled as a uniform rod, and the cables as light inextensible strings.

Given that the tension in the cable at C is twice the tension in the cable at A, find

a the tension in the cable at A,

b show that $AB = 120$ cm.

A small load of weight W newtons is attached to the girder at B. The load is modelled as a particle. The girder remains in equilibrium in a horizontal position. The tension in the cable at C is now three times the tension in the cable at A.

c Find the value of W.

29

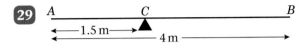

A uniform rod AB has mass 8 kg and length 4 m. A particle of mass 4 kg is attached to the rod at A and a particle of mass M kg is attached to the rod at B. The rod is supported at the point C, where $AC = 1.5$ m, and rests in equilibrium in a horizontal position, as shown in the diagram.

a Find the value of M.

b State how you used the information that the rod is uniform. **E**

30

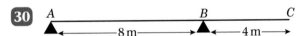

A uniform plank ABC, of length 12 m and mass 30 kg, is supported in a horizontal position at the points A and B, where $AB = 8$ m and $BC = 4$ m, as shown in the diagram. A woman of mass 60 kg stands on the plank at a distance of 2 m from A, and a rock of mass M kg is placed on the plank at the end C. The plank remains in equilibrium. The plank is modelled as a uniform rod, and the woman and the rock as particles.

Given that the forces exerted by the supports on the plank at A and B are equal in magnitude,

a find **i** the value of M, **ii** the magnitude of the force exerted by the support at A on the plank.

b State how you used the modelling assumption that the rock is a particle. **E**

31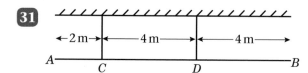

A light rod AB has length 10 m. It is suspended by two light vertical cables attached to the rod at the points C and D, where $AC = 2$ m, $CD = 4$ m, and $DB = 4$ m, as shown in the diagram. A load of weight 60 N is attached to the rod at A and a load of weight X N is attached to the rod at B. The rod is hanging in equilibrium in a horizontal position. Find, in terms of X,

a the tension in the cable at C,

b the tension in the cable at D.

c Hence show that $15 \leqslant X \leqslant 90$.

If the tension in either cable exceeds 120 N that cable breaks.

d Find the maximum possible value of X. **E**

32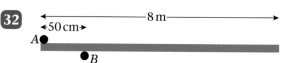

A large uniform plank of wood of length 8 m and mass 30 kg is held in equilibrium by two small steel rollers A and B, ready to be pushed into a saw-mill. The centres of the rollers are 50 cm apart. One end of the plank presses against roller A from underneath, and the plank rests on top of roller B, as shown in the diagram. The rollers are adjusted so that the plank remains horizontal and the force exerted on the plank by each roller is vertical.

a Suggest a suitable model for the plank to determine the forces exerted by the rollers.

b Find the magnitude of the force exerted on the plank by the roller at B.

c Find the magnitude of the force exerted on the plank by the roller at A. **E**

33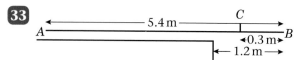

A plank of wood AB has length 5.4 m. It lies on a horizontal platform, with 1.2 m projecting over the edge, as shown in the diagram. When a girl of mass 50 kg stands at the point C on the plank, where $BC = 0.3$ m, the plank is on the point of tilting. By modelling the plank as a uniform rod and the girl as a particle,

a find the mass of the plank.

The girl places a rock on the end of the plank at A. By modelling the rock also as a particle,

b find, to two significant figures, the smallest mass of the rock which will enable the girl to stand on the plank at B without tilting.

c State briefly how you have used the modelling assumption that

  i the plank is uniform,

  ii the rock is a particle. **E**

34 Three forces $\mathbf{F}_1$, $\mathbf{F}_2$ and $\mathbf{F}_3$ act on a particle.

$\mathbf{F}_1 = (-3\mathbf{i} + 7\mathbf{j})\,\text{N}$, $\mathbf{F}_2 = (\mathbf{i} - \mathbf{j})\,\text{N}$,
$\mathbf{F}_3 = (p\mathbf{i} + q\mathbf{j})\,\text{N}$

a Given that the particle is in equilibrium, determine the value of p and the value of q.

The resultant of the forces $\mathbf{F}_1$ and $\mathbf{F}_2$ is $\mathbf{R}$.

b Calculate, in N, the magnitude of $\mathbf{R}$.

c Calculate to the nearest degree, the angle between the line of action of $\mathbf{R}$ and the vector $\mathbf{j}$. **E**

35

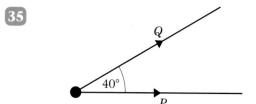

Two forces $\mathbf{P}$ and $\mathbf{Q}$, act on a particle. The force $\mathbf{P}$ has magnitude 5 N and the force $\mathbf{Q}$

has magnitude 3 N. The angle between the directions of **P** and **Q** is 40°, as shown in the diagram. The resultant of **P** and **Q** is **F**.

a Find, to three significant figures, the magnitude of **F**.

b Find, in degrees to one decimal place, the angle between the directions of **F** and **P**. **E**

36 Two forces **P** and **Q** act on a particle. The force **P** has magnitude 7 N and acts due north. The resultant of **P** and **Q** is a force of magnitude 10 N acting in a direction with bearing 120°. Find

a the magnitude of **Q**,

b the direction of **Q**, giving your answer as a bearing. **E**

37 Two forces $F_1 = (2\mathbf{i} + 3\mathbf{j})$ N and $F_2 = (\lambda\mathbf{i} + \mu\mathbf{j})$ N, where λ and μ are scalars, act on a particle. The resultant of the two forces is **R**, where **R** is parallel to the vector $\mathbf{i} + 2\mathbf{j}$.

a Find, to the nearest degree, the acute angle between the line of action of **R** and the vector **i**.

b Show that $2\lambda - \mu + 1 = 0$.

Given that the direction of F_2 is parallel to **j**,

c find, to three significant figures, the magnitude of **R**. **E**

38 A force **R** acts on a particle, where **R** = $(7\mathbf{i} + 16\mathbf{j})$ N. Calculate

a the magnitude of **R**, giving your answers to one decimal place,

b the angle between the line of action of **R** and **i**, giving your answer to the nearest degree

The force **R** is the resultant of two forces **P** and **Q**. The line of action of **P** is parallel to

the vector $(\mathbf{i} + 4\mathbf{j})$ and the line of action of **Q** is parallel to the vector $(\mathbf{i} + \mathbf{j})$.

c Determine the forces **P** and **Q** expressing each in terms of **i** and **j**. **E**

39 A particle P moves in a straight line with constant velocity. Initially P is at the point A with position vector $(2\mathbf{i} - \mathbf{j})$ m relative to a fixed origin O, and 2s later it is at the point B with position vector $(6\mathbf{i} + \mathbf{j})$ m.

a Find the velocity of P.

b Find, in degrees to one decimal place, the size of the angle between the direction of motion of P and the vector **i**.

Three seconds after it passes B the particle P reaches the point C.

c Find, in metres to one decimal place, the distance OC. **E**

40 A particle P is moving with constant velocity $(5\mathbf{i} - 3\mathbf{j})$ m s^{-1}. At time $t = 0$, its position vector, with respect to a fixed origin O, is $(-2\mathbf{i} + \mathbf{j})$ m. Find, to three significant figures,

a the speed of P,

b the distance of P from O when $t = 2s$. **E**

41 A boat B is moving with constant velocity. At noon, B is at the point with position vector $(3\mathbf{i} - 4\mathbf{j})$ km with respect to a fixed origin O. At 1430 on the same day, B is at the point with position vector $(8\mathbf{i} + 11\mathbf{j})$ km.

a Find the velocity of B, giving your answer in the form $p\mathbf{i} + q\mathbf{j}$.

At time t hours after noon, the position vector of B is **b** km.

b Find, in terms of t, an expression for **b**.

Another boat C is also moving with constant velocity. The position vector of C, c km, at time t hours after noon, is given by

$$\mathbf{c} = (-9\mathbf{i} + 20\mathbf{j}) + t(6\mathbf{i} + \lambda\mathbf{j}),$$
where λ is a constant.

Given that C intercepts B,

c find the value of λ,

d show that, before C intercepts B, the boats are moving with the same speed.

$\boxed{E}$

42 A particle P of mass 2 kg is moving under the action of a constant force $\mathbf{F}$ newtons. When $t = 0$, P has velocity $(3\mathbf{i} + 2\mathbf{j}) \, \mathrm{m\,s^{-1}}$ and at time $t = 4 \, \mathrm{s}$, P has velocity $(15\mathbf{i} - 4\mathbf{j}) \, \mathrm{m\,s^{-1}}$. Find

a the acceleration of P in terms of $\mathbf{i}$ and $\mathbf{j}$,

b the magnitude of $\mathbf{F}$,

c the velocity of P at time $t = 6 \, \mathrm{s}$. $\boxed{E}$

43 A ship S is moving with constant velocity $(-2.5\mathbf{i} + 6\mathbf{j}) \, \mathrm{km\,h^{-1}}$. At time 1200, the position vector of S relative to a fixed origin O is $(16\mathbf{i} + 5\mathbf{j}) \, \mathrm{km}$. Find

a the speed of S,

b the bearing on which S is moving.

The ship is heading directly towards a submerged rock R. A radar tracking station calculates that, if S continues on the same course with the same speed, it will hit R at the time 1500.

c Find the position vector of R.

The tracking station warns the ship's captain of the situation. The captain maintains S on its course with the same speed until the time is 1400. He then changes course so that S moves due north at a constant speed of 5 km h^{-1}. Assuming that S continues to move with this new constant vector, find

d an expression for the position vector of the ship t hours after 1400.

e The time when S will be due east of R.

f The distance of S from R at the time 1600.

In Questions 44 and 45 the horizontal unit vectors $\mathbf{i}$ and $\mathbf{j}$ are due east and due north respectively.

44 A model boat A moves on a lake with constant velocity $(-\mathbf{i} + 6\mathbf{j}) \, \mathrm{m\,s^{-1}}$. At time $t = 0$, A is at the point with position vector $(2\mathbf{i} - 10\mathbf{j}) \, \mathrm{m}$. Find

a the speed of A,

b the direction in which A is moving, giving your answer as a bearing.

At time $t = 0$, a second boat B is at the point with position vector $(-26\mathbf{i} + 4\mathbf{j}) \, \mathrm{m}$.

Given that the velocity of B is $(3\mathbf{i} + 4\mathbf{j}) \, \mathrm{m\,s^{-1}}$,

c show that A and B will collide at a point P and find the position vector of P.

Given instead that B has speed 8 m s^{-1} and moves in the direction of the vector $(3\mathbf{i} + 4\mathbf{j})$.

d find the distance of B from P when $t = 7 \, \mathrm{s}$. $\boxed{E}$

45 At time $t = 0$, a football player kicks a ball from the point A with position vector $(2\mathbf{i} + \mathbf{j}) \, \mathrm{m}$ on a horizontal football field. The motion of the ball is modelled as that of a particle moving horizontally with constant velocity $(5\mathbf{i} + 8\mathbf{j}) \, \mathrm{m\,s^{-1}}$. Find

a the speed of the ball,

b the position vector of the ball after t seconds.

The point B on the field has position vector $(10\mathbf{i} + 7\mathbf{j}) \, \mathrm{m}$.

c Find the time when the ball is due north of B.

At time $t = 0$, another player starts running due north from B and moves with constant speed $v\,\text{m s}^{-1}$. Given that he intercepts the ball,

d find the value of v.

e State one physical factor, other than air resistance, which would be needed in a refinement of the model of the ball's motion to make the model more realistic. **(E)**

46 A destroyer is moving due west at a constant speed of $10\,\text{km h}^{-1}$. It has radar on board which, at time $t = 0$, identifies a cruiser, $50\,\text{km}$ due west and moving due north with a constant speed of $20\,\text{km h}^{-1}$. The unit vectors $\mathbf{i}$ and $\mathbf{j}$ are directed due east and north respectively, and the origin O is taken to be the initial position of the destroyer. Each vessel maintains its constant velocity.

a Write down the velocity of each vessel in vector form.

b Find the position vector of each vessel at time t hours.

c Show that the distance $d\,\text{km}$ between the vessels at time t hours is given by

$$d^2 = 500t^2 - 1000t + 2500$$

The radar on the cruiser detects vessels only up to a distance of $40\,\text{km}$. By finding the minimum value of d^2, or otherwise,

d determine whether the destroyer will be detected by the cruiser's radar. **(E)**

47 In this question the vectors $\mathbf{i}$ and $\mathbf{j}$ are horizontal unit vectors in the directions due east and due north respectively. Two boats A and B are moving with constant velocities. Boat A moves with velocity $9\mathbf{j}\,\text{km h}^{-1}$. Boat B moves with velocity $(3\mathbf{i} + 5\mathbf{j})\,\text{km h}^{-1}$.

a Find the bearing on which B is moving.

At noon A is at the point O and B is $10\,\text{km}$ due west of O. At time t hours after noon, the position vectors of A and B relative to O are $\mathbf{a}\,\text{km}$ and $\mathbf{b}\,\text{km}$ respectively.

b Find expressions for $\mathbf{a}$ and $\mathbf{b}$ in terms of t, giving your answer in the form

$$p\mathbf{i} + q\mathbf{j}$$

c Find the time when B is due south of A.

At time time t hours after noon, the distance between A and B is $d\,\text{km}$. By finding an expression for $\overrightarrow{AB}$,

d show that $d^2 = 25t^2 - 60t + 100$.

At noon the boats are $10\,\text{km}$ apart.

e Find the time after noon at which the boats are again $10\,\text{km}$ apart. **(E)**

Examination style paper

1 A particle P of mass 0.2 kg is moving along a straight horizontal line with constant speed 12 m s^{-1}. Another particle Q of mass 0.8 kg is moving in the same direction as P, along the same straight horizontal line, with constant speed 2 m s^{-1}. The particles collide. Immediately after the collision, Q is moving with speed 6 m s^{-1}.

    a Find the speed of P immediately after the collision. (3)

    b State whether or not the direction of motion of P is changed by the collision. (1)

    c Find the magnitude of the impulse exerted on Q in the collision. (2)

                                                  Total: 6 marks

2 A particle P of weight W newtons is attached to one end of a light inextensible string. The other end of the string is attached to a fixed point O. The string is taut and makes an angle 30° with the vertical. The particle P is held in equilibrium under gravity by a force of magnitude 12 N acting in a direction perpendicular to the string, as shown. Find

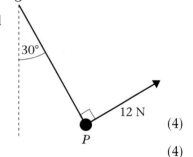

    a the tension in the string, (4)

    b the value of W. (4)

                                                  Total: 8 marks

3 A car is moving along a straight horizontal road. At time $t = 0$, the car passes a sign A with speed 8 m s^{-1} and this speed is maintained for 6 s. The car then accelerates uniformly from 8 m s^{-1} to 12 m s^{-1} in 9 s. The speed of 12 m s^{-1} is then maintained until the car passes a second sign B. The distance between A and B is 390 m.

    a Sketch a speed–time graph to illustrate the motion of the car as it travels from A to B. (2)

    b Find the time the car takes to travel from A to B. (5)

    c Sketch a distance–time graph to illustrate the motion of the car as it travels from A to B. (2)

                                                    Total: 9 marks

4 Two particles A and B are connected by a light inextensible string which passes over a fixed smooth pulley. The mass of A is 4 kg and the mass of B is m, where $m > 4$ kg. The system is released from rest with the string taut and the hanging parts of the string vertical, as shown. After release, the tension in the string is $\frac{1}{4}mg$.

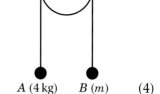

A (4 kg) B (m)

  a Find the magnitude of the acceleration of the particles. (4)

  b Find m. (4)

  c State how you have used the fact that the string is inextensible. (1)

Total: 9 marks

5 A particle of mass 0.8 kg is moving under the action of two forces $\mathbf{P}$ and $\mathbf{Q}$. The force $\mathbf{P}$ has magnitude 4 N and the force $\mathbf{Q}$ has magnitude 6 N. The angle between $\mathbf{P}$ and $\mathbf{Q}$ is 110°, as shown. The resultant of $\mathbf{P}$ and $\mathbf{Q}$ is $\mathbf{F}$. Find

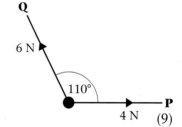

  a the angle between the direction of $\mathbf{F}$ and the direction of $\mathbf{P}$.

  b the magnitude of the acceleration of the particle. (9)

Total: 9 marks

6 A small stone, S, of mass 0.3 kg, slides with constant acceleration down a line of greatest slope of a rough plane, which is inclined at 30° to the horizontal. The stone passes through a point X with speed 1.5 m s^{-1}. Three seconds later it passes through a point Y, where $XY = 6$ m, as shown. Find.

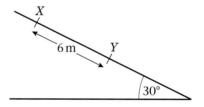

  a the acceleration of S, (2)

  b the magnitude of the normal reaction of the plane on S, (2)

  c the coefficient of friction between S and the plane. (5)

Total: 9 marks

7 *In this question the unit vectors* $\mathbf{i}$ *and* $\mathbf{j}$ *are due east and north respectively and position vectors are given with respect to a fixed origin* O.

A ship S is moving with constant velocity $(2\mathbf{i} - 3\mathbf{j})$ km h^{-1} and a ship R is moving with constant velocity $6\mathbf{i}$ km h^{-1}.

  a Find the bearing along which S is moving. (3)

At noon S is at the point with position vector $8\mathbf{i}$ km and R is at O. At time t hours after noon, the position vectors of S and R are $\mathbf{s}$ km and $\mathbf{r}$ km respectively.

  b Find $\mathbf{s}$ and $\mathbf{r}$, in terms of t. (2)

At time T hours, R is due north-east of S. Find

c the value of T, (4)

d the distance between S and R at time T hours. (3)

Total: 12 marks

8

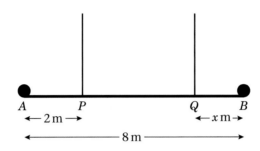

A uniform steel girder AB has length 8 m and weight 400 N. A load of weight 200 N is attached to the girder at A and a load of weight W newtons is attached to the girder at B. The girder and the loads hang in equilibrium, with the girder horizontal. The girder is held in equilibrium by two cables attached to the girder at P and Q, where $AP = 2$ m and $QB = x$ m, as shown. The girder is modelled as a uniform rod, the loads as particles and the cables as light inextensible strings.

a Show that the tension in the cable at Q is $\left(\dfrac{400 + 6W}{6 - x}\right)$ N. (5)

Given that the tension in the cable attached at P is five times the tension in the cable attached to Q,

b find W in terms of x, (6)

c deduce that $x < 2$. (2)

Total: 13 marks

Total: 75 marks

Answers

Exercise 2A

1 $20\,\text{m s}^{-1}$
2 $1.6\,\text{m s}^{-2}$
3 $0.625\,\text{m s}^{-2}$
4 $26\,\text{m}$
5 $20\,\text{m s}^{-1}$
6 $6\,\text{m s}^{-1}$ in direction $\overrightarrow{XY}$
7 a $9\,\text{m s}^{-1}$ **b** $72\,\text{m}$
8 a $3\,\text{m s}^{-1}$ **b** $\frac{1}{3}\,\text{m s}^{-2}$
9 a $9.2\,\text{m s}^{-1}$ **b** $33.6\,\text{m}$
10 a $18\,\text{km h}^{-1}$ **b** $312.5\,\text{m}$
11 a $8\,\text{s}$ **b** $128\,\text{m}$
12 a $0.4\,\text{m s}^{-2}$ **b** $320\,\text{m}$
13 a $0.25\,\text{m s}^{-2}$ **b** $16\,\text{s}$ **c** $234\,\text{m}$
14 a $19\,\text{m s}^{-1}$ **b** $2.4\,\text{m s}^{-2}$ **c** $430\,\text{m}$
15 a $x = 0.25$ **b** $150\,\text{m}$
16 b $500\,\text{m}$

Exercise 2B

1 $7\,\text{m s}^{-1}$
2 $\frac{2}{3}\,\text{s}^{-2}$
3 $2\,\text{m s}^{-2}$
4 $8.5\,\text{m s}^{-1}$
5 $2.5\,\text{s}$
6 $0.175\,\text{m s}^{-2}$
7 a $2.5\,\text{m s}^{-2}$ **b** $4.8\,\text{s}$
8 a $3.5\,\text{m s}^{-1}$ **b** $15.5\,\text{m s}^{-1}$
9 a $54\,\text{m}$ **b** $6\,\text{s}$
10 a $90\,\text{m}$ **b** $8.49\,\text{m s}^{-1}$ (3 s.f.)
11 a $3.3\,\text{s}$ (1 d.p.)   **b** $16.2\,\text{m s}^{-1}$ (1 d.p.)
12 a $4, 8$
    b $t = 4$: $4\,\text{m s}^{-1}$ in direction $\overrightarrow{AB}$, $t = 8$: $4\,\text{m s}^{-1}$ in direction $\overrightarrow{BA}$
13 a $0.8, 4$ **b** $15.0\,\text{m s}^{-1}$ (3 s.f.)
14 a $2\,\text{s}$ **b** $4\,\text{m}$
15 a $0.34\,\text{m s}^{-1}$ **b** $25.5\,\text{s}$ (3 s.f.)
16 a P: $(4t + t^2)\,\text{m}$ Q: $[3(t - 1) + 1.8(t - 1)^2]\,\text{m}$
    b $t = 6$ **c** $60\,\text{m}$

Exercise 2C

1 $10\,\text{m}$
2 $3.2\,\text{s}$ (2 s.f.)
3 $1.8\,\text{m}$ (2 s.f.)
4 $4.1\,\text{s}$ (2 s.f.)
5 $41\,\text{m}$ (2 s.f.)
6 a $29\,\text{m}$ (2 s.f.)   **b** $2.4\,\text{s}$ (2 s.f.)
7 a $5.5\,\text{m s}^{-1}$ (2 s.f.)   **b** $20\,\text{m s}^{-1}$ (2 s.f.)
8 a $40\,\text{m s}^{-1}$ (2 s.f.)   **b** $3.7\,\text{s}$ (2 s.f.)

9 a $39\,\text{m s}^{-1}$ **b** $78\,\text{m}$ (2 s.f.)
10 $4.7\,\text{m}$ (2 s.f.)
11 a $3.4\,\text{s}$ (2 s.f.)   **b** $29\,\text{m}$ (2 s.f.)
12 $2.8\,\text{s}$ (2 s.f.)
13 a 29 (2 s.f.)   **b** $6\,\text{s}$
14 $30\,\text{m}$ (2 s.f.)
15 a $5.6\,\text{m}$ (2 s.f.)   **b** $3.1\,\text{m}$ (2 s.f.)
18 a $1.4\,\text{s}$ (2 s.f.)   **b** $7.2\,\text{m}$ (2 s.f.)

Exercise 2D

1 a $2.25\,\text{m s}^{-2}$ **b** $90\,\text{m}$
2 a

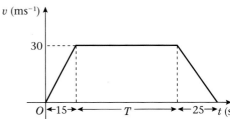

    b $360\,\text{m}$
3 a $0.4\,\text{m s}^{-2}$ **b** $\frac{8}{15}\,\text{m s}^{-2}$ **c** $460\,\text{m}$
4 a

    b $2125\,\text{m}$
5 a

    b $100\,\text{s}$
6 a $0.8\,\text{m s}^{-2}$ **b** $1960\,\text{m}$

7 a

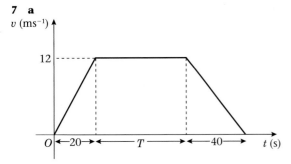

b $T = 320$ **c** $3840\,\text{m}$

8 a

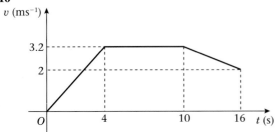

b $60\,\text{s}$

c

a (ms⁻²) graph

9 a $\frac{10}{3}$ **b** $\frac{20}{9}\,\text{ms}^{-2}$

10

v (ms⁻¹) graph with points 3.2, 2, at times 4, 10, 16

11 a

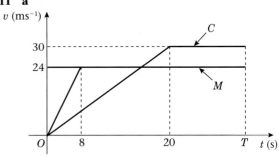

b $720\,\text{m}$

12 a

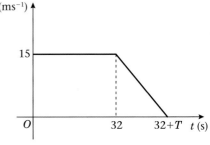

b $T = 12$

c

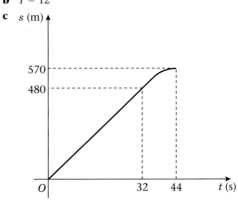

Exercise 2E

1 $2\,\text{m s}^{-2}$

2 $1.9\,\text{s}$ (2 s.f.)

3 $u = 8$

4 a $23\,\text{m}$ (2 s.f.)   **b** $2.1\,\text{s}$ (2 s.f.)

5 a $28\,\text{m s}^{-1}$ **b** $208\,\text{m}$

6 $0.165\,\text{m s}^{-2}$ (3 d.p.)

7 a

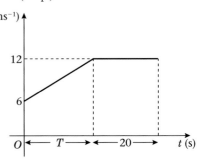

b $315\,\text{m}$ **c** $30\,\text{s}$

8 a $4.1\,\text{s}$ (2 s.f.)   **b** $40\,\text{m s}^{-1}$ (2 s.f.)

c air resistance

9 **a** $8\,\text{m}\,\text{s}^{-1}$ **b** $1.25\,\text{m}\,\text{s}^{-2}$ **c** $204.8\,\text{m}$

10 **a** $33\,\text{m}\,\text{s}^{-1}$ (2 s.f.)   **b** $3.4\,\text{s}$ (2 s.f.)

11 **a** $60\,\text{m}$ **b** $100\,\text{m}$

12 **a**

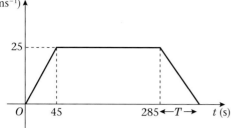

b $\frac{5}{7}\,\text{m}\,\text{s}^{-2}$

c

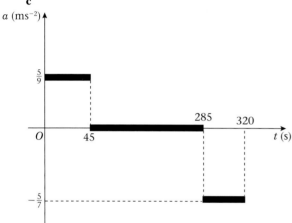

13 **a** $u = 11$ **b** $22\,\text{m}$

14 **a**

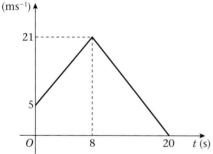

b $230\,\text{m}$

c

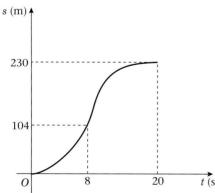

15 $1.2\,\text{s}$ (2 s.f.)

16 **a** $50\,\text{s}$ **b** $24.2\,\text{m}\,\text{s}^{-1}$ (3 s.f.)

17 $h = 39$ (2 s.f.)

18 **a** $32\,\text{m}\,\text{s}^{-1}$ **b** $90\,\text{m}$ **c** $5\,\text{s}$

19 **a**

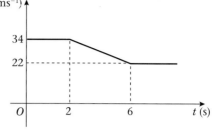

b $180\,\text{m}$

20 **a**

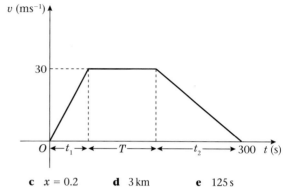

c $x = 0.2$ **d** $3\,\text{km}$ **e** $125\,\text{s}$

Exercise 3A

1 $39.2\,\text{N}$ **2** $50\,\text{kg}$

3 $112\,\text{N}$ **4** $4.2\,\text{N}$

5 $0.3\,\text{m}\,\text{s}^{-2}$ **6** $25\,\text{kg}$

7 **a** $25.6\,\text{N}$ **b** $41.2\,\text{N}$ **c** P is $34\,\text{N}$, Q is $49\,\text{N}$

8 **a** $2.1\,\text{kg}$ (2 s.f.)   **b** $1.7\,\text{kg}$ (2 s.f.)
   c $0.22\,\text{kg}$ (2 s.f.)

9 **a** $5.8\,\text{m}\,\text{s}^{-2}$ **b** $2.7\,\text{m}\,\text{s}^{-2}$ **c** $2.7\,\text{m}\,\text{s}^{-2}$

10 **a** $31.2\,\text{N}$ **b** $39.2\,\text{N}$ **c** $41.2\,\text{N}$

Exercise 3B

1 $2.3\,\text{N}$ (2 s.f.)

2 $0.35\,\text{N}$

3 **a** $0.9\,\text{m}\,\text{s}^{-2}$ **b** $7120\,\text{N}$ **c** $8560\,\text{N}$

4 $2.25\,\text{N}$

5 **a** $0.5\,\text{m}\,\text{s}^{-2}$ **b** $45\,\text{N}$

6 **a** $4\,\text{m}\,\text{s}^{-2}$ **b** $800\,\text{N}$

7 **a** $708\,\text{N}$ **b** $498\,\text{N}$
   c Her perception of her weight is the reaction force
         that she feels from the floor of the lift.

8 **a** $32\,\text{s}$ **b** $256\,\text{m}$
   c Air resistance unlikely to be constant.

9 **a** $1.5\,\text{m}\,\text{s}^{-2}$ **b** $60\,\text{kg}$ **c** $40\,\text{kg}$

10 **a** $2.9\,\text{m}$ (2 s.f.)   **b** $3.6\,\text{m}\,\text{s}^{-1}$ (2 s.f.)
   c $2.17\,\text{s}$ (3 s.f.)

Exercise 3C

1 **a** **i** $11.3\,\text{N}$ (3 s.f.)   **ii** $4.10\,\text{N}$ (3 s.f.)
   b **i** $0\,\text{N}$ **ii** $-5\,\text{N}$
   c **i** $-5.14\,\text{N}$ (3 s.f.)   **ii** $6.13\,\text{N}$ (3 s.f.)
   d **i** $8.66\,\text{N}$ (3 s.f.)   **ii** $-5\,\text{N}$
   e **i** $-3.86\,\text{N}$ (3 s.f.)   **ii** $-4.60\,\text{N}$ (3 s.f.)
   f **i** $F\cos\theta\,\text{N}$ **ii** $F\sin\theta\,\text{N}$

2 a i -2 N   **ii** 6.93 N (3 s.f.)
   b i 8.13 N (3 s.f.)   **ii** 10.3 N (3 s.f.)
   c i $P\cos\alpha + Q - R\sin\beta$ **ii** $P\sin\alpha - R\cos\beta$

Exercise 3D

1 a i 3 N
     ii $F = 3$ N and body remains at rest
   b i 7 N
     ii $F = 7$ N and body remains at rest
   c i 7 N
     ii $F = 7$ N and body accelerates
     iii $1\,\mathrm{m\,s^{-2}}$
   d i 6 N
     ii $F = 6$ N and body remains at rest
   e i 9 N
     ii $F = 9$ N and body remains at rest in limiting equilibrium
   f i 9 N
     ii $F = 9$ N and body accelerates
     iii $0.6\,\mathrm{m\,s^{-2}}$
   g i 3 N
     ii $F = 3$ N and body remains at rest
   h i 5 N
     ii $F = 5$ N and body remains at rest in limiting equilibrium
   i i 5 N
     ii $F = 5$ N and body accelerates
     iii $0.2\,\mathrm{m\,s^{-2}}$
   j i 6 N
     ii $F = 6$ N and body accelerates
     iii $1.22\,\mathrm{m\,s^{-2}}$ (3 s.f.)
   k i 5 N
     ii $F = 5$ N and body accelerates
     iii $3.85\,\mathrm{m\,s^{-2}}$ (3 s.f.)
   l i 12.7 N (3 s.f.)
     ii The body accelerates.
     iii $5.39\,\mathrm{m\,s^{-2}}$ (3 s.f.)
2 a $R = 88$ N, $\mu = 0.083$ (2 s.f.)
   b $R = 80.679$ N, $\mu = 0.062$ (2 s.f.)
   c $R = 118$ N, $\mu = 0.13$ (2 s.f.)

Exercise 3E

1 $3.35\,\mathrm{m\,s^{-2}}$ (3 s.f.)
2 a 27.7 N (3 s.f.)   **b** $2.12\,\mathrm{m\,s^{-2}}$
3 a $2.43\,\mathrm{m\,s^{-2}}$ (3 s.f.)   **b** $4.93\,\mathrm{m\,s^{-1}}$ (3 s.f.)
4 28 N
5 0.20 (2 s.f.)
6 0.15 (2 s.f.)
7 a 88.8 N (3 s.f.)   **b** 0.24 (2 s.f.)
8 a $\dfrac{13g}{15}$ **b** 23.5 m (3 s.f.)
   c 2.35 s (3 s.f.)   **d** $12.4\,\mathrm{m\,s^{-1}}$ (3 s.f.)

Exercise 3F

1 a 4 N   **b** 0.8 N
2 a $R = 45$ **b** 100 N
3 a $3\,\mathrm{m\,s^{-2}}$ **b** 2500 N
4 a 33.6 N (3 s.f.)   **b** $2\frac{2}{7}$ m
5 a $0.613\,\mathrm{m\,s^{-2}}$ (3 s.f.)   **b** 27.6 N (3 s.f.)
   c 39.0 N (3 s.f.)
6 $2.8\,\mathrm{m\,s^{-1}}$

7 a $0.569\,\mathrm{m\,s^{-2}}$ (3 s.f.)   **b** $0.56\,mg$
8 a $1.12\,\mathrm{m\,s^{-2}}$ **b** 4100 N
9 a 21.9 N   **b** 0.418 (3 s.f.)   **c** 38 N (2 s.f.)
10 a $2\,\mathrm{m\,s^{-2}}$ **b** 600 N   **c** 100 m

Exercise 3G

1 $30\,\mathrm{m\,s^{-1}}$
2 $2.5\,\mathrm{m\,s^{-1}}$
3 2.59 N s
4 $6.5\,\mathrm{m\,s^{-1}}$
5 $3\,\mathrm{m\,s^{-1}}$

Exercise 3H

1 $4\,\mathrm{m\,s^{-1}}$
2 $2\frac{2}{9}\,\mathrm{m\,s^{-1}}$
3 $4.5\,\mathrm{m\,s^{-1}}$
4 a $2\frac{2}{3}\,\mathrm{m\,s^{-1}}$ **b** $2\frac{2}{3}$ N s
5 a $1\,\mathrm{m\,s^{-1}}$ and direction unchanged
   b 15 N s
6 10
7 a $\dfrac{2u}{3}$; direction reversed
   b $8\,mu$
8 Larger $8\,\mathrm{m\,s^{-1}}$ and smaller $4\,\mathrm{m\,s^{-1}}$
9 a 3 **b** $\dfrac{9mu}{2}$
10 a $3\,\mathrm{m\,s^{-1}}$ **b** 4.5
11 a $4\,\mathrm{m\,s^{-1}}$ in same direction
   b $3\,\mathrm{m\,s^{-1}}$ in opposite direction
12 a $3\,\mathrm{m\,s^{-1}}$ **b** 6 kg

Exercise 3I

1 a 0.103 kg   **b** 4.103 kg
2 0.14 (2 s.f.)
3 a $\frac{1}{2}u = v$ **b** $6mu$
4 a 0.22 m (2 s.f.)   **b** $\dfrac{14g}{25}$
   c $1.1\,\mathrm{m\,s^{-1}}$ (2 s.f.)
5 0.12 (2 s.f.)
6 a 9.8 N   **b** 9.8 N
7 a $14\,\mathrm{m\,s^{-1}}$ **b** $\frac{35}{3}\,\mathrm{m\,s^{-1}}$ **c** 0.75 m (2 s.f.)
8 a $\frac{1}{3}g$ **b** $3.6\,\mathrm{m\,s^{-1}}$ (2 s.f.)   **c** $2\frac{2}{3}$ m
   d i acceleration both masses equal
     ii same tension in string either side of pulley
9 a 540 N   **b** 180 N   **c** 450 N
10 1000 N vertically downwards
11 a 2000 **b** 36 m
12 a $1.75\,\mathrm{m\,s^{-1}}$ **b** 0.45 N s
13 a $2.5\,\mathrm{m\,s^{-1}}$ **b** $15\,000$ N s
14 a $0.7\,\mathrm{m\,s^{-1}}$ **b** unchanged   **c** 8.25 N s
15 $\frac{4}{5}$
16 a $1.25\,\mathrm{m\,s^{-1}}$ **b** 0.77 (2 s.f.)
17 0.44 (2 s.f.)
18 a 1.3 N (2 s.f.)   **b** 19 m (2 s.f.)
19 $\frac{5}{28}$
20 a 830 N (2 s.f.)   **b** 1500 N (2 s.f.)
   c 1700 N (2 s.f.)
   d Air resistance would reduce speed of lift as it falls and so impulse would be reduced.
21 a 18 N (2 s.f.)   **b** $0.12\,\mathrm{m\,s^{-2}}$ (2 s.f.)
22 a 243 N (3 s.f.)   **b** $3.08\,\mathrm{m\,s^{-2}}$ (3 s.f.)
   c 36.7 m (3 s.f.)

23 **a** $7.5\,\mathrm{m\,s^{-1}}$ **b** $11\,000$ (2 s.f.)
    c R could be modelled as varying with speed.
24 **a** $\frac{12}{7}g\,\mathrm{N}$ **b** 1.2
25 **a** $3.2\,\mathrm{m\,s^{-2}}$ **b** $5.3\,\mathrm{N}$ (2 s.f.)
    c 0.75 (2 s.f.)
    d The information that the string is inextenible has been used when, in part **c** the acceleration of A has been taken as equal to the acceleration of B obtained in part **a**.
26 **a** $18\,\mathrm{N}$ (2 s.f.)   **b** 2
    c $4.2\,\mathrm{N\,s}$ **d** $\frac{2}{7}\,\mathrm{s}$

Review Exercise 1

Exercise A
1 **a** $1.12\,\mathrm{m\,s^{-2}}$ **b** $31.25\,\mathrm{s}$
2 **a** $3.6\,\mathrm{m\,s^{-2}}$ **b** $AC = 760\,\mathrm{m}\;BC = 440\,\mathrm{m}$
3 **a** 14.4 **b** $36\,\mathrm{m\,s^{-1}}$
4 **a** $0.5\,\mathrm{m\,s^{-2}}$ **b** $7.5\,\mathrm{m\,s^{-1}}$
5 $\dfrac{AB}{BC} = \dfrac{51}{40}$
6 $108\,\mathrm{m\,s^{-1}}$ (3 s.f.)
7 **a** 24 **b** $OA = 96\,\mathrm{m}$
    c $4\,\mathrm{s}$ and $12\,\mathrm{s}$
8 **a** $2.5\,\mathrm{m\,s^{-2}}$ **b** $31.7\,\mathrm{m\,s^{-1}}$ (3 s.f.)
    c $1.69\,\mathrm{s}$ (3 s.f.)
9 **a** $6t - t^2$ **b** $7\,\mathrm{m}$ **c** $t = 5$
10 **a** 34 (2 s.f.)   **b** $60\,\mathrm{m}$ (2 s.f.)
11 **a** $28\,\mathrm{m\,s^{-1}}$ **b** $5.7\,\mathrm{s}$ (2 s.f.)
12 2 or 4
13 **a** 14 (2 s.f.)   **b** $23\,\mathrm{m\,s^{-1}}$ (2 s.f.)
14 $10\,\mathrm{m}$ (2 s.f.)
15 **a** 28 **b** $4\frac{2}{7}\,\mathrm{s}$
16 **a**

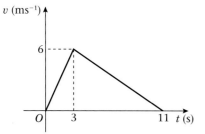

    b $33\,\mathrm{m}$
17 **a**

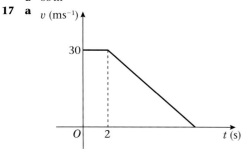

    b $18\,\mathrm{s}$
18 **a** constant acceleration   **b** constant speed
    c $30.5\,\mathrm{m}$

19 **a**

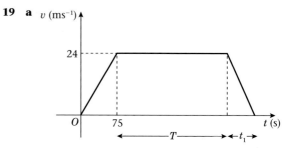

    b $0.48\,\mathrm{m\,s^{-2}}$ **c** 250 **d** $375\,\mathrm{s}$
20 **a**

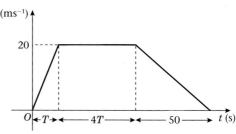

    b 8 **c** $2.5\,\mathrm{m\,s^{-2}}$
21 **a** $162\,\mathrm{m}$ **b** 6.2 **c** $0.56\,\mathrm{m\,s^{-2}}$
22 **a** $185\,\mathrm{s}$ **b** $2480\,\mathrm{m}$
    c

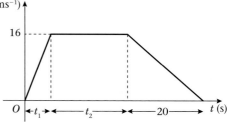

23 **a**

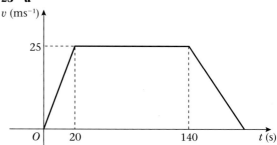

    b $200\,\mathrm{s}$ **c** $60\,\mathrm{s}$ **d** $50\,\mathrm{m\,s^{-1}}$
24 **a**

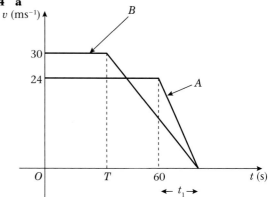

    b $10\,\mathrm{s}$ **c** 34

25 a v (ms^{-1}) ... 10, 3, O, 12, 27, t (s)

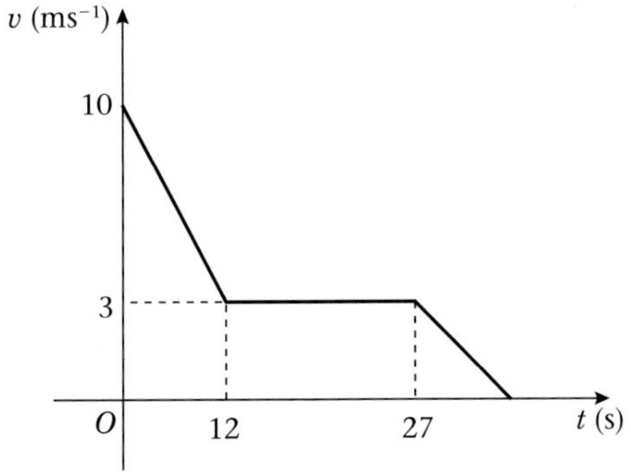

b 78 m **c** 35 s

d a (ms^{-2}) ... 12, 27, 35, O, t (s), $-\frac{3}{8}$, $-\frac{7}{12}$

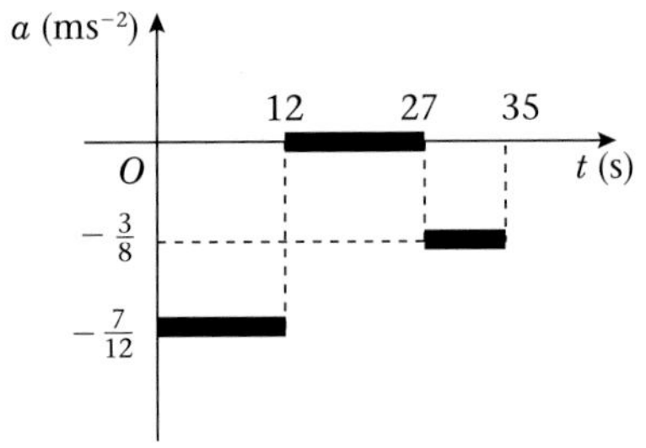

26 a v (ms^{-1}) ... 50, 10, O, 80, 140, 220, t (s)

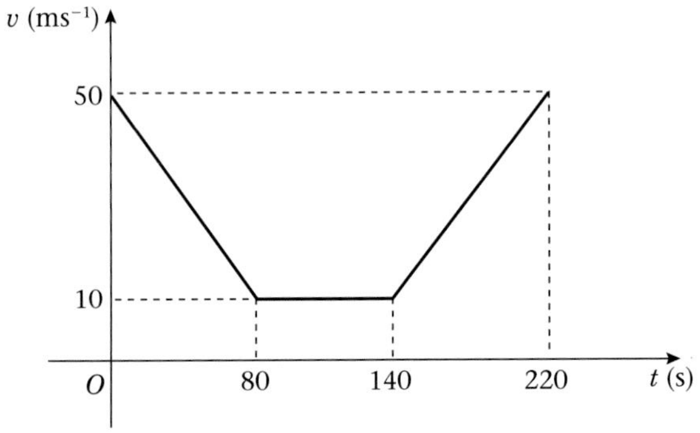

b 5400 m **c** 112 s

27 a bus has not overtaken cyclist

b v (ms^{-1}) ... bus, cycle, 12, 8, O, 6, T, t (s)

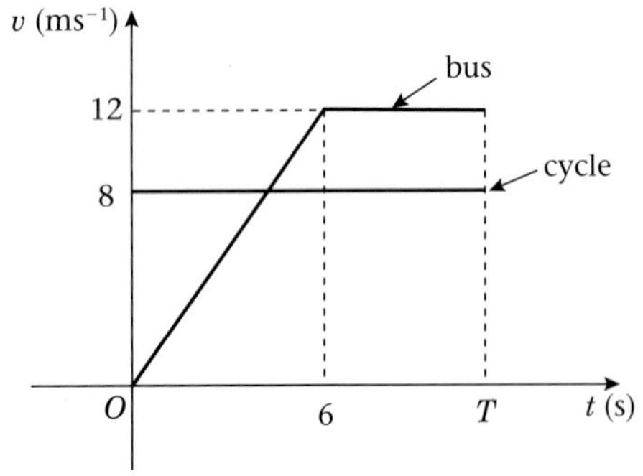

28 a v (ms^{-1}) ... 60, 30, Train A, Train B, O, 12, 40, 64, T, t (s)

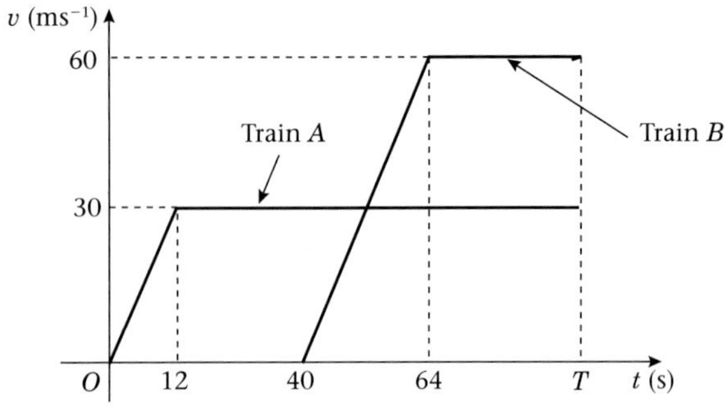

b 98

29 a v (ms^{-1}) ... 15, $\frac{3}{4}T$, O, t_1, T, t_2, t (s)

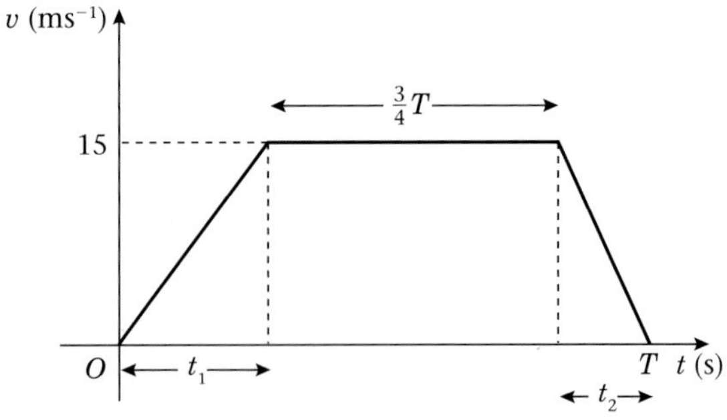

b 96 s **c** $\frac{15}{16}$ m s^{-2}

30 a v (ms^{-1}) ... 30, 10, O, 240, t (s), $\leftarrow t \rightarrow \leftarrow 240 - 4t \rightarrow \leftarrow 3t \rightarrow$, $\leftarrow T \rightarrow$

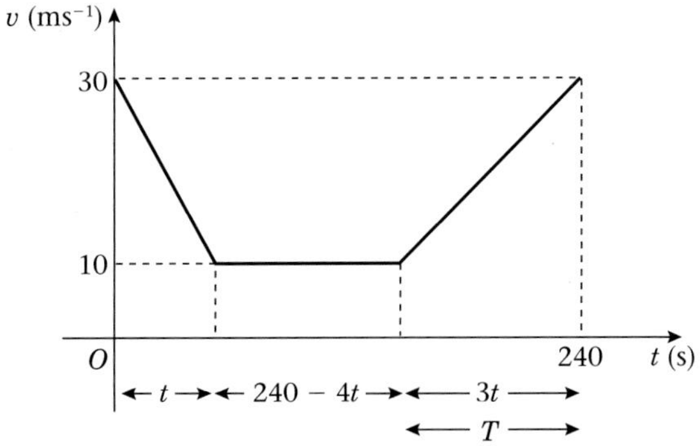

b $\frac{1}{6}$ **c** 800 m

31 $66\frac{2}{3}$ m s^{-1}

32 a 13 m s^{-1} **b** 2 m s^{-1} in direction $\overrightarrow{CB}$

33 6.3 N

34 a 2.25 m s^{-1} direction of motion unchanged

b 1.5 N s

35 a 2.4 m s^{-1} **b** due west **c** 3000 kg

36 a A 2.2 m s^{-1} B 3 m s^{-1}

b 0.4 N s **c** 1.6 N s

37 a 3 m s^{-1} **b i** $m = 3.6$ **ii** 18 N s

38 750 N

39 a 0.42 N **b** 2.5

40 a 2.45 m s^{-2} **b** 0.25

41 0.30 (2 s.f.)

42 0.37 (2 s.f.)

43 a 3 m s^{-2} **b** 14.8 m s^{-1} (3 s.f.)

c 0.1 kg **d** 3.06 s (3 s.f.)

44 a 8.6 m s^{-1} **b** 24 m **c** 79.2 m

45 520 (2 s.f.)

46 a 0.693 m s^{-2} (3 s.f.) **b** 7430 N (3 s.f.)

c 28 kN (2 s.f.)

47 **a** $3.6\,\mathrm{m\,s^{-2}}$ **b** 0.75 (2 s.f.)   **c** $14\,\mathrm{m}$ (2 s.f.)
48 **a** 0.35 (2 s.f.)
  b normal reaction unchanged hence friction force unchanged
  c $1500\,\mathrm{N}$ (2 s.f.)
49 **a** $15\,\mathrm{m\,s^{-1}}$ **b** 991 (3 s.f.)
50 **a** 22.4 **b** 4.64 (3 s.f.)   **c** 6380 (3 s.f.)
  d Consider air resistance due to motion under gravity
51 **a** $4.2\,\mathrm{m\,s^{-2}}$ **b** $3.4\,\mathrm{N}$ (2 s.f.)
  c $2.9\,\mathrm{m\,s^{-1}}$ (2 s.f.)   **d** $0.69\,\mathrm{s}$ (2 s.f.)
52 **a** $1.4\,\mathrm{m\,s^{-2}}$ **b** $3.4\,\mathrm{N}$ (2 s.f.), $4.2\,\mathrm{N}$
53 **a** **i** $1050\,\mathrm{N}$ **ii** $390\,\mathrm{N}$
  b $3\,\mathrm{m\,s^{-2}}$
54 **a** $2.2\,\mathrm{m\,s^{-2}}$ (2 s.f.)   **b** $22\,\mathrm{N}$ (2 s.f.)
  c $4.4\,\mathrm{m}$ (2 s.f.)
55 **a** $\frac{6}{5}mg$ **b** 0.693 (3 s.f.)
  c $\frac{6}{5}mg$ vertically downwards
56 **a** $1.2\,\mathrm{m\,s^{-2}}$ **b** $16\,\mathrm{N}$
  c The information that the string is inextensible has been used in assuming that the accelerations of P and Q, and hence of the whole system, are the same.
  d $3\,\mathrm{s}$ **e** $20\,\mathrm{m\,s^{-1}}$
57 **a** $1.0\,\mathrm{m}$ (2 s.f.)   **b** $17\,\mathrm{N}$ (2 s.f.)
  c $26\,\mathrm{N}$, direction bisecting angle ABC

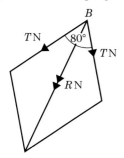

  d 0.55 (2 s.f.)
58 **a** $11\,500\,\mathrm{N}$ **b** $6.2\,\mathrm{m\,s^{-2}}$
  c $3700\,\mathrm{N}$ **d** $31\,\mathrm{m\,s^{-1}}$ (2 s.f.)
59 **a** $0.24\,\mathrm{m\,s^{-2}}$ **b** $530\,\mathrm{N}$ (2 s.f.)   **c** $54\,\mathrm{m}$
  d normal reaction of the road on the car is increased when the tow bar breaks

Exercise 4A

1 **a** $Q - 5\cos 30° = 0$ **b** $P - 5\sin 30° = 0$
  c $Q = 4.33\,\mathrm{N}$ $P = 2.5\,\mathrm{N}$
2 **a** $Q - P\cos 60° = 0$ **b** $P\sin 60° - 4\sqrt{3} = 0$
  c $Q = 4\,\mathrm{N}$ $P = 8\,\mathrm{N}$
3 **a** $9 - P\cos 30° = 0$ **b** $Q + P\sin 30° - 8 = 0$
  c $Q = 2.80\,\mathrm{N}$ $P = 10.4\,\mathrm{N}$
4 **a** $9 - P\cos 30° = 0$ **b** $Q + P\sin 30° - 8 = 0$
  c $Q = 2.80\,\mathrm{N}$ $P = 10.4\,\mathrm{N}$
5 **a** $4\cos 45° + P\cos\theta - 7 = 0$
  b $4\sin 45° - P\sin\theta = 0$ **c** $\theta = 34.1°$ $P = 5.04\,\mathrm{N}$
6 **a** $6\cos 45° - 2\cos 60° - P\sin\theta = 0$
  b $6\sin 45° + 2\sin 60° - P\cos\theta - 4 = 0$
  c $\theta = 58.7°$ $P = 3.80\,\mathrm{N}$
7 **a** $P\cos\theta + 8\sin 40° - 7\cos 35° = 0$
  b $P\sin\theta + 7\sin 35° - 8\cos 40° = 0$
  c $\theta = 74.4°$ (allow 74.3°) $P = 2.20$ (allow 2.19)
8 **a** $9\cos 40° + 3 - P\cos\theta - 8\sin 20° = 0$
  b $P\sin\theta + 9\sin 40° - 8\cos 20° = 0$
  c $\theta = 13.6°$ $P = 7.36$

9 **a** $P\cos 30° - Q\cos 45° - 8\cos 45° = 0$
  b $P\sin 30° + Q\sin 45° - 8\sin 45° - 4 = 0$
  c $P = 11.2$ (3 s.f.) $Q = 5.73$ (3 s.f.)
10 **a** $Q\cos 60° - P\cos 60° + 5\sin 45° - 6\sin 45° = 0$
  b $P\sin 60° + Q\sin 60° - 5\cos 45° - 6\cos 45° = 0$
  c $P = 3.784\,\mathrm{N}$ $Q = 5.198\,\mathrm{N}$
11 **a** $Q - 10\sin 45° = 0$
  b $P - 10\cos 45° = 0$
  c $P = 7.07\,\mathrm{N}$ $Q = 7.07\,\mathrm{N}$
12 **a** $Q + 2\cos 60° - 6\sin 60° = 0$
  b $P - 2\sin 60° - 6\cos 60° = 0$
  c $P = 4.73$ $Q = 4.20$
13 **a** $8\sin 30° - Q\cos 30° = 0$
  b $P - Q\sin 30° - 8\cos 30° = 0$
  c $P = 9.24\,\mathrm{N}$ $Q = 4.62\,\mathrm{N}$
14 **a** $8\cos 45° - 10\sin 30° - Q° = 0$
  b $P + 8\sin 45° - 10\cos 30° = 0$
  c $P = 3.00\,\mathrm{N}$ $Q = 0.657\,\mathrm{N}$
15 **a** $2 + 8\sin 30° - P\cos\theta = 0$
  b $4 - 8\cos 30° + P\sin\theta = 0$
  c $\theta = 26.0°$ $P = 6.68\,\mathrm{N}$

Exercise 4B

1 $34.7\,\mathrm{N}$
2 **a** $20\,\mathrm{N}$ **b** 1.77
3 14.4
4 $S = 30.4$ or 30.5, $T = 43.0$
5 **a** $5.46\,\mathrm{N}$ **b** $0.762\,\mathrm{kg}$
6 **a** 1.46 **b** $55\,\mathrm{g}$
7 **a** $3\,\mathrm{N}$ **b** $2\,\mathrm{N}$
8 **a** 2.6 **b** 4.4
9 **a** $F = 19.6m$, $R = 9.8m$
  b $F = 17m$ (3 s.f.), $R = 0$
  c $P = 11.2$ (3 s.f.), $Q = 5.73$ (3 s.f.)
10 $13.9\,\mathrm{N}$
11 39.2
12 $37.2\,\mathrm{N}$ (3 s.f.)
13 $F = 12.25$, $R = 46.6$ (3 s.f.)
14 $P = 20.4$ (3 s.f), $R = 0.400$

Exercise 4C

1 0.446
2 0.123
3 **a** $1.5\,\mathrm{N}$ **b** not limiting
4 **a** 40
  b The assumption is that the crate and books may be modelled as a particle.
5 **a** 11.9 **b** 6.40
6 0.601 (accept 0.6)
7 **a** 13.3 **b** $F = 3.33$, $X = 9.54$
8 **a** $9.97\,\mathrm{N}$ down the plane
  b $-2.7\,\mathrm{N}$
  c $\mu \geqslant 0.439$
9 **a and b** $X = 44.8$ (accept 44.7), $R = 51.3$
10 $F = 22.1$, $T = 102$ (3 s.f.)
11 **a** $T = 3.87$ **b** $T = 2.75$
12 0.758

Exercise 4D

1 $\alpha = 52.6°$, $T = 24.7$
2 **a and b** The weight of the particle is $80\,\mathrm{N}$ and the tension in the second string is $69.3\,\mathrm{N}$ (3 s.f.).
3 **a** 6.93 (3 s.f.)   **b** 3.46 (3 s.f.)

4 a 43° (to nearest degree)
  b 53 N (to nearest Newton)
5 a 138.2° (1 d.p.)   **b** 8.95 (2 d.p.)
6 $T = 17.3$, $S = 21.3$
7 $R = 20.7$, $\mu = 0.24$ (2 s.f.)
8 $\mu = 0.296$ (3 s.f.)
9 363
10 11°
11 a 0.577
  b The book was modelled as a particle.
12 a $W = 11.4$ **b** $R = 13.9$
13 2.2 (1 d.p.)
14 0.75 (2 d.p.)
15 0.262 (3 s.f.)
16 a

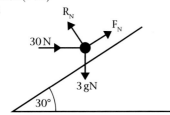

  b 40.46   **c** 0.279 (3 s.f.)
17 a $R = 88.3$ **b** $P = 74.7$
  c resultant force 9 N down plane and box will move
18 11.0
19 $\dfrac{\mu \cos \alpha - \sin \alpha}{\cos \alpha - \mu \sin \alpha}$
20 a 15.7 (3 s.f.)   **b** 0.625
21 0.577 (3 s.f.)
22 0.399 (3 s.f.)
23 a 0.684 (3 s.f.)   **b** 2.33
  c As 2.425 > 1.596 the ring is not in equilibrium.

Exercise 5A

1 6 Nm clockwise
2 10.5 Nm clockwise
3 13 Nm anticlockwise
4 0 Nm
5 10 Nm anticlockwise
6 11.6 Nm clockwise
7 30.5 Nm anticlockwise
8 0 Nm
9 13.3 Nm clockwise
10 33.8 Nm anticlockwise

Exercise 5B

1 a 5 Nm anticlockwise   **b** 13 Nm clockwise
  c 19 Nm anticlockwise   **d** 11 Nm anticlockwise
  e 4 Nm clockwise   **f** 7 Nm anticlockwise
2 a 16 Nm clockwise   **b** 1 Nm anticlockwise
  c 10 Nm clockwise   **d** 7 Nm clockwise
  e 0.5 Nm anticlockwise   **f** 9.59 Nm anticlockwise

Exercise 5C

1 a 10 N, 10 N   **b** 15 N, 5 N
  c 8.6 N, 11.4 N   **d** 12.6 N, 7.4 N
2 a 7.5, 17.5   **b** 30, 35
  c 245, $2\frac{2}{3}$ **d** 49, 1.5
3 0.5 m from B
4 59 N
5 31 cm from the broomhead

6 16.25 N, 13.75 N
7 1.71 m
8 5
9 $\frac{2}{3}$ m
10 2.05 m
11 a 15 N   **b** rod will tilt   **c** 3.17 m

Exercise 5D

1 2.4 N, 3.6 N
2 3.5 m from A
3 $\frac{1}{3}$ m from A
4 a 29.4 N, 118 N   **b** 4.25 m

Exercise 5E

1 a 105 N   **b** 140 N   **c** 1.03 m
2 b $0 \leqslant x < \frac{7}{4}$
3 a $40g$ **b** $x = \frac{1}{2}$
  c i the weight acts at the centre of the plank
    ii the plank remains straight
    iii the man's weight acts at a single point
4 b $W = 790 - 300x$ **c** $x = 2.53$, $W = 30$
5 a 200 N   **b** 21 cm
6 a 36 kg   **b** 2.2 m
7 a 19.6 N   **b** 5
8 a 588 N   **b** $\frac{2}{3}$ m
9 a 125 N   **b** 1.8 m

Exercise 6A

1 8.60 km from starting point on bearing of 054°
2 10 km, 7.2 km on bearing of 326°
3 7.43 km, 062°
4 9.13 km, 340°
5 31.8 km, 261°
6 171.4°, 328.6°
7 3.01 km, 220°

Exercise 6B

1 a $2\mathbf{b}$ **b** $\mathbf{d}$ **c** $\mathbf{b}$
  d $2\mathbf{b}$ **e** $\mathbf{d} + \mathbf{b}$ **f** $\mathbf{d} + \mathbf{b}$
  g $-2\mathbf{d}$ **h** $-\mathbf{b}$ **i** $2\mathbf{d} + \mathbf{b}$
  j $-\mathbf{b} + 2\mathbf{d}$ **k** $-\mathbf{b} + \mathbf{d}$ **l** $-\mathbf{b} - \mathbf{d}$
2 a $2\mathbf{m}$ **b** $2\mathbf{p}$ **c** $\mathbf{m}$
  d $\mathbf{m}$ **e** $\mathbf{p} + \mathbf{m}$ **f** $\mathbf{p} + \mathbf{m}$
  g $\mathbf{p} + 2\mathbf{m}$ **h** $\mathbf{p} - \mathbf{m}$ **i** $-\mathbf{m} - \mathbf{p}$
  j $-2\mathbf{m} + \mathbf{p}$ **k** $-2\mathbf{p} + \mathbf{m}$ **l** $-\mathbf{m} - 2\mathbf{p}$
3 a $2\mathbf{p}$ **b** $2\mathbf{r}$ **c** $-2\mathbf{p} + 2\mathbf{r}$
  d $-\mathbf{p} + \mathbf{r}$ **e** $\mathbf{p} + \mathbf{r}$ **f** $\mathbf{r}$
  g $-\mathbf{p}$ **h** $-2\mathbf{r} + \mathbf{p}$
4 $\frac{2}{3}\mathbf{a} + \frac{1}{3}\mathbf{b}$
5 $\frac{3}{5}\mathbf{a} + \frac{2}{5}\mathbf{b}$

Exercise 6C

1 $4\mathbf{i}$ **2** $5\mathbf{i} + 2\mathbf{j}$ **3** $-3\mathbf{i} + \mathbf{j}$
4 $2\mathbf{i} + 3\mathbf{j}$ **5** $-2\mathbf{i} - \mathbf{j}$ **6** $-3\mathbf{j}$

Exercise 6D

1 a $6\mathbf{i} + 2\mathbf{j}$ **b** $10\mathbf{i} + 8\mathbf{j}$ **c** $7\mathbf{j}$
  d $10\mathbf{i} + \mathbf{j}$ **e** $-2\mathbf{i} + \mathbf{j}$ **f** $-2\mathbf{i} - 10\mathbf{j}$
  g $14\mathbf{i} - 7\mathbf{j}$ **h** -8 **i** $+9\mathbf{j}$

2 a 5   **b** 10
  c 13   **d** 4.47 (3 s.f.)
  e 5.83 (3 s.f.)   **f** 8.06 (3 s.f.)
  g 5.83 (3 s.f.)   **h** 4.12 (3 s.f.)

3 a 53.1° above   **b** 53.1° below
  c 67.4° above   **d** 63.4° above

4 a 149° to the right   **b** 29.7° to the right
  c 31.0° to the left   **d** 104° to the left

5 a $\lambda = 5$ **b** $\mu = -\frac{3}{2}$

6 a $\lambda = \frac{1}{3}$ **b** $\mu = -1$

  c $s = -1$ **d** $t = -\frac{1}{17}$

7 a 3.61 (3 s.f.), 023°   **b** 4.12 (3 s.f.), 104°
  c 3.61 (3 s.f.), 304°   **d** 2.24 (3 s.f.), 243°

Exercise 6E

1 a $5\,\text{m s}^{-1}$ **b** $25\,\text{km h}^{-1}$
  c $5.39\,\text{m s}^{-1}$ **d** $8.06\,\text{cm s}^{-1}$

2 a $50\,\text{km}^{1}$ **b** 51.0 m
  c 4.74 km   **d** 967 cm

3 a $5\,\text{m s}^{-1}$, 75 m   **b** $5.39\,\text{m s}^{-1}$, 16.2 m
  c $5.39\,\text{km h}^{-1}$, 16.2 km   **d** $13\,\text{km h}^{-1}$, 6.5 km

Exercise 6F

1 a $8\mathbf{i} + 3\mathbf{j}$ **b** $2\mathbf{i} - 7\mathbf{j}$
  c $-17\mathbf{i} + 16\mathbf{j}$ **d** $7\mathbf{i} - 13\mathbf{j}$

2 a $2\mathbf{i} + 5\mathbf{j}$ **b** $\mathbf{i} + 3\mathbf{j}$
  c $2\mathbf{i} + 4\mathbf{j}$ **d** $2\mathbf{i} - 5\mathbf{j}$
  e $-2\mathbf{i} - 5\mathbf{j}$

3 a $6\mathbf{i} - 8\mathbf{j}$ **b** $-12\mathbf{i} + 9\mathbf{j}$
  c $-4.5\mathbf{i} + 6\mathbf{j}$ **d** $5\mathbf{i} + 5\mathbf{j}$
  e $-4\mathbf{i} + 6\mathbf{j}$ **f** $3\sqrt{2}\,\mathbf{i} - 5\sqrt{2}\,\mathbf{j}$
  g $-4\sqrt{3}\,\mathbf{i} - 2\sqrt{3}\,\mathbf{j}$ **h** $-3\sqrt{5}\,\mathbf{i} + 6\sqrt{5}\,\mathbf{j}$

4 a $6\mathbf{i} + 12\mathbf{j}$ **b** $-7\mathbf{i} + 4\mathbf{j}$
  c $-2\mathbf{i} + 6\mathbf{j}$ **d** $10\mathbf{i} - 13\mathbf{j}$
  e $2\mathbf{i} - 3\mathbf{j}$,   **f** $4.61\,\text{m s}^{-1}$
  g 4   **h** 2.5

5 a $5\mathbf{i} + 12\mathbf{j}$, $13\,\text{m s}^{-1}$ **b** $6\mathbf{i} - 5\mathbf{j}$, $7.81\,\text{m s}^{-1}$
  c $-2\mathbf{i} - 5\mathbf{j}$, $5.39\,\text{m s}^{-1}$ **d** $-3\mathbf{i} - 2\mathbf{j}$, $3.61\,\text{m s}^{-1}$
  e $7\mathbf{i} + 9\mathbf{j}$, $11.4\,\text{m s}^{-1}$

6 $4.8\mathbf{i} - 6.4\mathbf{j}$

7 10.1 m

8 $2.03\,\text{m s}^{-1}$

9 a $2t\mathbf{i} + (-500 + 3t)\mathbf{j}$ **b** 721 m

10 a $7t\mathbf{i} + (400 + 7t)\mathbf{j}$, $(500 - 3t)\mathbf{i} + 15t\mathbf{j}$
  b $350\mathbf{i} + 750\mathbf{j}$

11 a $(1 + 2t)\mathbf{i} + (3 - t)\mathbf{j}$, $(5 - t)\mathbf{i} + (-2 + 4t)\mathbf{j}$
  b 5.39 km

12 a $121\,\text{m s}^{-1}$, $6.08\,\text{m s}^{-1}$ **b** $18\mathbf{i} - 3\mathbf{j}$
  c $15\mathbf{i} - 12\mathbf{j}$

Exercise 6G

1 a $\mathbf{i} - 8\mathbf{j}$ **b** $-5\mathbf{i} + \mathbf{j}$
  c $2\mathbf{i} + 5\mathbf{j}$ **d** $-3\mathbf{i} + 2\mathbf{j}$

2 a 8.06, 82.9° below   **b** 5.10, 169° above
  c 5.39, 68.2° above   **d** 3.61, 146° above

3 a $6\mathbf{i}$, $3\mathbf{i}\,\text{m s}^{-2}$ **b** $3\mathbf{i} - 2\mathbf{j}$, $(\mathbf{i} - \frac{2}{3}\mathbf{j})\,\text{m s}^{-2}$
  c $3\mathbf{i} - 2\mathbf{j}$, $(\frac{3}{4}\mathbf{i} - \frac{1}{2}\mathbf{j})\,\text{m s}^{-2}$ **d** $\mathbf{i} - 6\mathbf{j}$, $(\frac{1}{2}\mathbf{i} - 3\mathbf{j})\,\text{m s}^{-2}$

4 a 5.83 N, 59°   **b** 6.32 N, 18.4°
  c 6.40 N, 38.7°

5 a 5.83 N, 3.83 N   **b** 4.39 N, 5.38 N
  c 4.20 N, 6.53 N   **d** 14.4 N, 12.7 N
  e 4.54 N, 31.9°

Exercise 6H

1 a $p = 2$, $q = -6$ **b** 6.32 N
  c 18°

2 a $(-3 + t)\mathbf{i} + (10 + t)\mathbf{j}$ **b** 4.24 km
  c 1630

3 a $\mathbf{p} = 6t\mathbf{i}$, $\mathbf{q} = (12 - 3t)\mathbf{i} + (6 + 6t)\mathbf{j}$
  b 38.4 km   **c** $1\frac{1}{3}$

4 a 3   **b** $10.2\,\text{m s}^{-2}$
  c 168.7°

5 a $-4\mathbf{i} + 2\mathbf{j}\,\text{m s}^{-2}$ **b** 22.4 N
  c 26 m

6 a 031°   **b** $\mathbf{a} = 6t\mathbf{i}$, $\mathbf{b} = 3t\mathbf{i} + (-10 + 5t)\mathbf{j}$
  c 1400   **d** 1456

7 a 108°   **b** $(-2 + 9t)\mathbf{i} + (-4 - 3t)\mathbf{j}$
  c 41, -23

8 a 124°   **b** $(3 - 2t)\mathbf{i} + (-2 + 3t)\mathbf{j}$
  c $11.2\,\text{m s}^{-1}$ **d** 1

9 a $9.85\,\text{m s}^{-1}$ **b** $(3 + 4t)\mathbf{i} + (2 + 9t)\mathbf{j}$
  c 6.5 s   **d** $7.46\,\text{m s}^{-1}$

10 a $(5\mathbf{i} + 3\mathbf{j})\,\text{km h}^{-1}$
  b $(10 + 5t)\mathbf{i} + (15 + 3t)\mathbf{j}$, $(-16 + 12t)\mathbf{i} + 26\mathbf{j}$
  c 0525

Review Exercise 2

1 a 48   **b** 41.6
2 a 40.8°   **b** 22.7 N (3 s.f.)
3 a 42.9°   **b** 52.8 N (3 s.f.)
4 a 35.1 N   **b** 33.0 N (3 s.f.)
5 a 26.1   **b** 51.4 (3 s.f.)
6 a 7.5   **b** 12
7 a $\dfrac{5mg}{2}$ **b** $\frac{4}{7}$
8 a 86.6   **b** 100
9 47.5 (1 d.p.)
10 a 19.9 N   **b** 3.46
11 a 23.0   **b** 17.6
  c The friction is not limiting and so equilibrium is maintained.
12 a 18.7   **b** 0.60 (2 s.f.)
  c Equilibrium is maintained and so the parcel does not move.
13 a 1.68   **b** 0.548
14 a 257 (3 s.f.)   **b** 12.5 s
15 a 131 N   **b** 209 N
  c i Friction acts down the slope,. magnitude 0.4R
    ii No acceleration so net force on package is zero
16 a 0.270   **b** $3.76\,\text{m s}^{-2}$ down the plane
17 a 109.2   **b** $1.46\,\text{m s}^{-2}$
18 5.6 m
19 a 88.2 N   **b** 0.875 m
20 $\frac{7}{8}$ m
21 a 2   **b** 0.6 m
22 a 911 N   **b** 1176 N   **c** 2.25 m
23 a 784 N   **b** 0.5 m
24 1.6d
25 a 50 N   **b** 1.9 m
26 a 0.75   **b** 24 N   **c** 144 N
  d The weight of the rock acts precisely at B.

27 a 1.25
 b The weight of the beam acts through its midpoint at C.
 c 0.4 m
28 a 70 N **b** 120 cm **c** 30
29 a 0.8
 b The weight acts through the mid-point of the rod.
30 a i 7.5 kg **ii** 477.75 N
 b Assumed that the centre of mass acts at the point C.
31 a $90 - X$ **b** $2X - 30$
 c $15 \leqslant X \leqslant 90$ **d** 75
32 a Model the plank as a uniform rod.
 b $240g$ **c** $210g$
33 a 30 kg **b** 3.6 kg
 c i plank is uniform so weight acts through mid-point
  ii rock is a particle so mass of rock acts through end-point A
34 a $p = 2, q = -6$ **b** $2\sqrt{10}$ or 6.32 (3 s.f.)
 c 18° (to nearest degree)
35 a 7.55 N **b** 14.8°
36 a 14.8 **b** 144.2°
37 a 63.4° **b** $2\lambda - \mu + 1 = 0$
 c 4.47 (3 s.f.)
38 a 17.5 (1 d.p.)° **b** 66°
 c $P = 3\mathbf{i} + 12\mathbf{j}$ $Q = 4\mathbf{i} + 4\mathbf{j}$
39 a $2\mathbf{i} + \mathbf{j}$ **b** 26.6° **c** 12.6 m
40 a 5.83 **b** 9.43
41 a $(2\mathbf{i} + 6\mathbf{j})$ km h^{-1} **b** $(3\mathbf{i} - 4\mathbf{j}) + (2\mathbf{i} + 6\mathbf{j})t$
 c $\lambda = -2$ **d** $\sqrt{40}$ km h^{-1}
42 a $3\mathbf{i} - 1.5\mathbf{j}$ **b** 6.71 **c** $21\mathbf{i} - 7\mathbf{j}$
43 a 6.5 km h^{-1} (2 s.f.) **b** 337°
 c $8.5\mathbf{i} + 23\mathbf{j}$ **d** $11\mathbf{i} + (17 + 5t)\mathbf{j}$
 e 1512 **f** 4.72 km
44 a 6.08 m s^{-1} **b** 3517°
 c $-5\mathbf{i} + 32\mathbf{j}$ **d** 21 m
45 a 9.43 m s^{-1} **b** $2\mathbf{i} + \mathbf{j} + t(5\mathbf{i} + 8\mathbf{j})$
 c 1.6 s **d** 4.25 m s^{-1}
 e friction on field – so velocity of ball not constant *or* vertical component of ball's motion *or* time for player to accelerate
46 a **velocities** destroyer: $-10\mathbf{i}$ km h^{-1}, cruiser: $20\mathbf{j}$ km h^{-1}
 b **position vectors** destroyer: $-10t\mathbf{i} = \mathbf{d}$ cruiser: $-50\mathbf{i} + 20t\mathbf{j} = \mathbf{c}$
 c $d^2 = 500t^2 - 1000t + 2500$
 d as $44.72 > 40$ cruiser will not be able to detect destroyer
47 a 031° (to nearest degree)
 b $(3t - 10)\mathbf{i} + 5t\mathbf{j}$ **c** 15.20
 d $d^2 = 25t^2 - 60t + 100$ **e** 14.24

Examination Style Paper

1 a 4 m s^{-1}
 b The direction of motion of P has been changed by the collision.
 c 3.2 N s.
2 a $12\sqrt{3}$ N. **b** 24
3 a

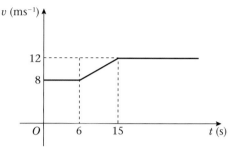

 b 36 s.
 c

s (m) graph: points 390, 138, 48 on s-axis; t (s) axis with 6, 15, 36

4 a $\frac{3}{4}g$
 b For A $m = 28$
 c The accelerations of the particles have the same magnitude.
5 a 70.9°, to 3 significant figures.
 b 7.46 m s^{-2}, to 3 significant figures.
6 a $\frac{1}{3}$ m s^{-2}
 b 2.5 N, to 2 significant figures.
 c 0.54, to 2 significant figures.
7 a 146
 b $\mathbf{s} = 8\mathbf{i} + (2\mathbf{i} - 3\mathbf{j})t$, $\mathbf{r} = 6t\mathbf{i}$
 c $T = 8$
 d $24\sqrt{2}$ km.
8 a Adequate working to show Q is $\left(\dfrac{400 + 6W}{6 - x}\right)$ N
 b $W = \dfrac{600(2 - x)}{30 + x}$

Index